Financial Engineering Explained

W0090650

Series Editor
Wim Schoutens
Department of Mathematics
Katholieke Universiteit Leuven,
Heverlee, Belgium

Financial Engineering Explained is a series of concise, practical guides to modern finance, focusing on key, technical areas of risk management and asset pricing. Written for practitioners, researchers and students, the series discusses a range of topics in a non-mathematical but highly intuitive way. Each self-contained volume is dedicated to a specific topic and offers a thorough introduction with all the necessary depth, but without too much technical ballast. Where applicable, theory is illustrated with real world examples, with special attention to the numerical implementation.

Series Advisory Board:
Peter Carr, Executive Director, NYU Mathematical Finance; Global Head of Market Modeling, Morgan Stanley.
Ernst Eberlein, Department of Mathematical Stochastics, University of Freiburg.
Matthias Scherer, Chair of Mathematical Finance, Technische Universität München.

Titles in the series:
Equity Derivatives Explained, Mohamed Bouzoubaa
The Greeks and Hedging Explained, Peter Leoni
Smile Pricing Explained, Peter Austing
Financial Engineering with Copulas Explained, Matthias Scherer and
    Jan-Frederik Mai
Interest Rates Explained Volume 1, Jörg Kienitz

Forthcoming titles:
Interest Rates Explained Volume 2, Jörg Kienitz
Submissions: Wim Schoutens - wim@schoutens.be

More information about this series at
http://www.springer.com/series/14984

Marc Henrard

Algorithmic Differentiation in Finance Explained

Marc Henrard
Advisory Partner,
Open Gamma
London, United Kingdom

ISBN 978-3-319-53978-2   ISBN 978-3-319-53979-9(eBook)
DOI 10.1007/978-3-319-53979-9

Library of Congress Control Number: 2017941212

Cover image © Rowan Moore / Getty Images

Printed on acid-free paper

This Palgrave Macmillan imprint is published by Springer Nature
The registered company is Springer International Publishing AG
The registered company address is: Gewerbestrasse 11, 6330 Cham, Switzerland

Contents

List of Figures

List of Tables

Preface

The book's title originally proposed to the editor was "Forty-nine shades of Algorithmic Differentiation"; a title that he duly refused as not fitting the series guideline very well and potentially distracting the target audience. The proposed title was not only a marketing exercise; there may not be as many as forty-nine variations of Algorithmic Differentiation (AD) presented in the book, nevertheless it is one of the goals of the book to present as many variations, shades or faces of the subject as possible, and not a one-size-fit-all approach. While exaggerating the number of shades, why not going to fifty? I'm expecting the readers to implement their own version of the technique, so for each reader it will be *"fifty shades of AD."* I also removed one from the number in the title to avoid confusion between AD and grey; AD is a very colorful subject!

Algorithmic Differentiation (AD) has been used in engineering and computer science for a long time. The term *Algorithmic Differentiation* can be explained as "the art of calculating the differentiation of functions with a computer."

When I first heard about it, it sounded to me like some kind of dark art or black magic where you get plenty of results from your computer for (almost) free. As a beginner in finance I was told that "there is no free lunch." Certainly this technique could not be applied to finance.

I nevertheless scrutinized the technique carefully, having heard about it several times from serious quants in serious conferences. What I discovered is that at the same time, it is not magic – it is mathematics –, it is not free – you have to invest in development time – and it appears like magic and (almost) free at run time.

Six years later, I have invested a lot of time understanding the technique and developing quantitative finance libraries which implement AD. Now I ask myself: "Why is not everybody in finance using AD?" I don't have an answer to this last question! But I'm very biased in this. I'm like a newly converted faithful to a religion, I don't understand how I was not a believer in the past and I don't understand how others don't believe.

It took me a while to write this book. Between my first article on the subject in 2010 and this book, more than 6 years have passed.

Why it took me so long?

Chacun sa méthode... Moi, je travaille en dormant et la solution de tous les problèmes, je la trouve en rêvant. (Personal translation: Each his own method... Myself, I work sleeping and the solution to all problems, I find it dreaming).

Drôle de drame (1937) – Marcel Carné

This means a lot of working nights to dream all those pages.

Obviously I cannot write something without referring to my previous book: *Interest Rate Modelling in the Multi-curve Framework: Foundations, Evolution, and Implementation* Henrard (2014). Most of the examples in this book are related to interest rate modeling and the multi-curve framework.

In my previous book I tried to follow the steps of giants in the art of relevant and irrelevant quotes. For this book, I changed the style. Each chapter starts with a couple of sentences summarizing some of the chapter's discoveries, like in old detective fictions. Those of the readers who know me personally, may have notice my zealous collection of (paper) books and the personal library to collect them. That collection contains a lot of detective fictions.

This book grew from seminars, lectures and training I presented on Algorithmic Differentiation. The first of those seminars was probably a "pizza seminar" at OpenGamma in March 2012. Pizza have disappeared from OpenGamma diet since, but seminars are still going on... Training are also continuing, at conference workshops or in-house in some banks, and I see more and more interest in the AD subject. Professional finance magazines, like *Risk* and *Structured Product*, have proposed articles on the subject – see for example Sherif (2015) and Davidson (2015) – and interviewed me for them.

This book has benefitted of the valuable feedback from numerous people in the financial "quant" community, among them Wim Schoetens, Luca Capriotti, Yupeng Jiang, and Andrea Macrina

In the book I use the term "we" with the general and vague meaning of the author, the finance community and the readers. The terms "I," "me" or "my" are used with the precise meaning of *the author personally with its opinions and biases*. The term "I" should be used as a warning sign that the sentence contains opinions and maybe not only facts. You have been warned!

The book was written astride 2014 and 2015. Among my resolutions for 2015 was to be more clear and direct in expressing my opinions. You may find some vapor of that resolution in the book. Some opinions may appear strong or lacking nuance. It is for clarity sake, not to be rude to others opinions. Even if I have spent some time in England, I have not learned the art of *understatement* yet!

Enjoy!

Marc Henrard
London, Brussels and Wépion – September 2016

Supplied with cleverness of every imaginable type,
Man ventures once towards evil, and then towards good.

Sophocles, Chorus in Antigone

Use Algorithmic Differentiation for the good!

Bibliography

Davidson, C. (2015). Structured products desks join the AAD revolution. *Structured Products*, 11(7):14–18.

Henrard, M. (2014). *Interest Rate Modelling in the Multi-curve Framework: Foundations, Evolution and Implementation*. Applied Quantitative Finance. London:Palgrave Macmillan. ISBN: 978-1-137-37465-3.

Sherif, N. (2015). Chips off the menu. *Risk Magazine*, 27(1):12–17.

La lecture des bons livres est comme une conversation avec les plus honnêtes gens des siècles passés qui en ont été les auteurs, et même une conversation étudiée, en laquelle ils ne nous découvrent que les meilleures de leurs pensées.

René Descartes
Discours de la méthode (1637)

Chapter 1
Introduction

Where the characters are introduced – Where the non-characters are released from duty – Where the adjoint come on stage – The code to the code.

1.1 Differentiation in Finance

In quantitative finance, computing the price of financial instruments is only part of the game; most of the (computer) time is spent in calculating the first order derivatives of the price with respect to the different inputs, the so-called deltas or *greeks*. This does not include only the simple greeks in a Black-Scholes formula, like Delta, Vega or Rho but more importantly, and more costly in computation time, the bucketed deltas to interest rate or credit curves and vega to interpolated volatility surfaces or cubes. The bucketed delta is usually more computationally costly as the multiple curves in a multi-curve framework are often made of 20–100 points and there are 20–100 deltas to compute.

The computation of function derivatives is not restricted to finance. The problem has important application in engineering in general. The Algorithmic Differentiation (AD) techniques have been used in different fields for a long time. The techniques comes in two modes: *standard*, *forward* or *tangent* and *adjoint* or *backward*. Both modes have their advantages and their drawbacks. The most useful one in finance is the adjoint mode; the reason for this is explained below.

In the computation of derivatives, two aspects have to be taken into account: precision and speed. The AD is an answer to both concerns; it is the proverbial stone with which two birds can be killed. With AD, the results are computed to machine precision level; the only approximations are the ones of processing doubles, there is no theoretical approximation. The speed is in general very good, with *all* derivatives and the function value computed in a time equivalent to the time required to compute 3–4 function values. This time is for *all* the derivatives of the function as one block, even if there are hundreds of them.

The algorithmic differentiation-like methods were popularized in finance by Giles and Glasserman (2006). The method used was not pure Algorithmic Differentiation method, and they are called in some places algebraic adjoint approaches. The method was applied to Monte Carlo computation for the Libor Market Model; those methods are usually computational time intensive. The method was extended to Bermudan style derivatives by Leclerc et al. (1999) and further extended in

© The Author(s) 2017
M. Henrard, *Algorithmic Differentiation in Finance Explained*,
Financial Engineering Explained, DOI 10.1007/978-3-319-53979-9_1

Denson and Joshi (2009a) and Denson and Joshi (2009b). Similar methods are used for Monte Carlo-based calibration in Kaebe et al. (2009).

The first uses of fully fledged Algorithmic Differentiation in finance, include the application to correlation risk in credit models in Capriotti and Giles (2010), to Monte Carlo credit risk in Capriotti et al. (2011), to gamma of CDO tranches in Joshi and Yang (2010). In two different papers, Capriotti (2011) and Homescu (2011) describe the AD general technique and its uses in finance. The review paper Capriotti and Giles (2012) describes several of those applications. More recent results also include application to second order greeks in Capriotti (2015), to PDE in Capriotti et al. (2015) and to least square Monte Carlo in Capriotti et al. (2016). The acronym "AAD" to describes Adjoint Algorithmic Differentiation was apparently first used in Capriotti (2011) and Capriotti and Giles (2012).

This book does not present all the computer science theoretical results regarding the method. For the theoretical results, we refer to Griewank and Walther (2008) and Naumann and du Toit (2014). Those references go a lot deeper on the computer science theoretical aspects of the technique.

In particular in this book, we will not prove theoretical results regarding the efficiency of the method. We will show the efficiency through practical examples and implementations in quantitative finance. The goal is a pragmatic implementation, from very simple cash flow discounting to model calibration, with some stops at Black-Scholes formula, SABR smile and multi-curve calibration.

The direct application of AD in finance is to the computation of first order risks. For that reason I will use the name of *Risk Manager* for the business end users of the libraries developed. It does not mean that the title Risk Manager is printed on their business card, but that they have an interest in the risks of the financial instruments. This category includes traders, quantitative analysts, risk managers, portfolio managers, model validators, senior management, supervisors, etc. With that definition in mind, the trader is probably the risk manager which has the most interest in AD as he probably requires speed more than others. On the other side I will call *developers* or computer scientists the people that write the code in the libraries. But again, their business card can indicate a completely different title, like quantitative analyst, model validator, risk manager, developer, architect, etc.

1.2 Standard and Adjoint: A Starter

AD comes in two modes: *standard*, *forward* or *tangent* and *adjoint* or *backward*. Those modes refer to the way the derivatives are computed.

The starting point of our presentation is always that the algorithm to compute the value of the function has been developed and implemented in some code. From there we want to write efficient ways to compute the derivatives of the value function with respect to the inputs. The design of the initial algorithm for the value itself is not discussed in this book.

The *standard* mode starts from the inputs and computes the derivatives with respect to the inputs going through the code in the same order as the value algorithm to finish with the derivatives of the output with respect to the inputs. This seems natural and would probably be the first implementation one would think of. If we first apply function g and then function f to the inputs a to get the output $z = f(g(a))$, the first approach to derivative computation is to compute $Dg(a)$, then $Df(g(a))$ and combine[1] them to get $D(f \circ g)(a)$. This approach put the emphasis on a: we want to compute the derivatives of everything with respect to a.

The other way to look at this problem is to view the problem as computing the derivative of the output. The output is z; obviously we known the derivative of z with respect to z, it is 1. Let go back one step and look at the previous operation, which is f. The function f is applied to an intermediary result, let's call it b for the moment. We have $z = f(b)$. We can compute the derivative of z with respect to b easily, there is only one step: $Dz = Df(b)$. But b itself is the result of a previous operation: $b = g(a)$. We are not finished yet, we still have to compose with $Dg(a)$ to obtain $Dz = Df(b).Dg(a)$.

Obviously the two approaches provide the same answer with the same computation complexity in the very simple example above with only one input and only one output. But what happens with multi-dimensional inputs and outputs? If you have one input and several outputs, the forward mode is easy: compute the derivative with respect to the unique input and for each line of code in the valuation algorithm add a line of code with the derivative. Several output for the value leads to several outputs for the derivatives, but the number of line of code is one to one between value and derivatives.

What if we have several inputs and one output? At each line of code, as we progress forward through the code, we have to add *several* line of code, one for each input. The computation time of the derivative algorithm will be of the order of magnitude of the number of inputs multiplied by the computation time of the value. In term of computational complexity it does not seems different from a one-side finite difference approach – the finite difference approach is presented in Section 1.5 –, there is one extra computation required for each input.

What is the more likely situation in finance? The standard computation is one present value (for each instrument) with the value depending of numerous inputs: interest rate curves, credit spreads, forex rates, etc. the usual problem is a "numerous inputs and one output" type of question. The usual finance case does not seems to be improved, from a computation time perspective, with respect to one-side finite difference by the use of forward AD.

In the case of the adjoint version, we start from the end, i.e. from the one dimensional present value output and at each step going backward, we ask what is the derivative of the (unique) output with respect to the intermediary value we are computing at this line. So for each line of code in the value algorithm, there is one new line of code in the derivatives algorithm. The code may be slightly more complex, but in term of order of magnitude it is the same for the value and all the derivatives.

---

[1] The mathematics required for the theoretical part of AD are described in Chapter 2. The notation used are described in Appendix A.

Now, this is a real game changer in finance. If we can compute the derivatives, all the derivatives – and this can be as many as 100 or 1,000 derivatives – in roughly the same time as the value, we are saving a huge computation time. Suppose we are computing the present value of a portfolio with 100 inputs (interest rate curves and forex) in one minute; if we want all the 100 sensitivities, just give us two minutes more. That is different from the standard answer: the portfolio present value is computed in one minute, if you want all the sensitivities, please come back in one hour. Risk management is not any more a lot more expansive in CPU term than simple present value computation.

The explanation above is an introduction to why we focus to the adjoint version of the algorithmic differentiation in this book. More details about the standard and adjoint modes will be given in the next chapter called "The principles of Algorithmic Differentiation."

1.3 Why Exact Derivative?

The goal of Algorithmic Differentiation is to compute the derivatives of a function, in the mathematical sense of the term, as precisely and fast as possible using a computer. A remark I have often heard in this context is the following: "I don't want to compute the derivative, I want the bump and recompute number with precisely one basis point shift to take convexity into account." Even if I understand the origin of the remark, I disagree strongly that it makes sense and it is relevant.

I understand that the exact derivatives multiplied by one basis point is not the same as an actual movement of one basis point. The derivative is the slope in the first order development and is just that, it is the precise slope of the first order development. If you want the exact value of function, the derivative will provide an approximation, an indication of the true result. Obviously if you know that the market will move by one basis point and exactly one basis point you don't want to compute the exact derivative, you want to compute the exact profit. But if you know that the market will move by exactly plus one basis point, what is the point of risk management? You want to compute the profit, fine, but not the sensitivity to the unknown movement of a quantity for which you just pretended to know its actual movement. Is the one basis point bump and recompute computation better than the exact derivative? Yes it is, when you know that the market will move by exactly plus one basis point. If the movement is minus one basis point, the result will be worst. So if you know that the movement is exactly plus one basis point, you have an easy way to get rich. Your problem is not to know what your risk is, exactly, it is "How rich I'm? Exactly!"

The same argument could be said for larger bumps where the argument is "one basis point does not really hurt my profit," only 5 basis point does, so I want a bump and recompute with 5 basis points. But 5 basis points in which direction? In the direction adverse to my portfolio. Ok, I see what you want now. Actually you have a number in mind; a currency amount that hurts you and you want to see if it can be reached by your position. What you are looking for is not sensitivity, it is VaR. Give me you probability tolerance, and I will give you a VaR for your portfolio and you can compare it with your hurting figure.

In this context, I have heard similar remarks than the one above but with a twist, like "I don't want to compute the derivative, I want the symmetrical bump and recompute number with precisely five basis point shift." The five basis point is to take into account the expected absolute shift of the market – in the same sense as in expected shortfall –, the symmetrical is to take into account the uncertainty of the direction of the market or "to be more precise." The symmetrical approach will make it more precise in the sense it is closer to the *exact derivative* but not that it is more precise to estimate your potential loss or gain. So again the proposed methodology is not in line with the stated goal.

From my point of view, I have never heard a convincing business argument for not computing the exact derivatives but an approximated bump and recompute ones. Maybe there is one, but you will need to find someone else – or an older and wiser me – to tell you about it.

1.4 Code in This Book

In the book, in several places some code is used as examples of the different concepts described. The code examples are written in Java. The main reason why I have used Java is that it is the programming language which I'm using on a daily basis and for which I have written quantitative finance libraries which are open source, giving reader the access to full size production grade libraries that implement the ideas described in the book. There is no other fundamental reason for it. Some may even say that this is not a very good choice and a language like C++, with operator overload, would be better, specially for Automatic Algorithmic Differentiation (see Chapter 4). I selected a language that, to my opinion, make it easy to transmit the ideas I want to describe and in which some quantitative finance libraries are written. There is talk about introducing operator overloading in Java, but it will have to wait for Java 10 at best. If this happen, I may revisit my code.

Moreover the code proposed in the book should be viewed as pseudo-code. All the important ingredients to make it work are there, but some details, like punctuation, programming language syntax, types may be slightly off when it shortens the code or make it easier to read. There is certainly no implicit claim that the pseudo-code will even compile as provided. If this warning saddens you, you may want to glance at Section 1.7.

1.5 What Is Not in This Book: Finite Difference

The most used approximation in finance for the computation of first order derivatives of scalar functions are the *forward difference*

$$Df(a) = \frac{f(a+\epsilon) - f(a)}{\epsilon} + O(\epsilon).$$

and *backward difference*

$$Df(a) = \frac{f(a) - f(a - \epsilon)}{\epsilon} + O(\epsilon).$$

These two methods are also known as one-sided *finite difference* or *bump and recompute*.

In term of computational complexity, there is one computation for the initial value, plus one similar computation for the value at each shifted input. This means a complexity of two for a one dimensional function and complexity $1 + n$ for a n-inputs/1-output function.

Take a very simple function, like $f(x) = \exp(x + x^2)$ and look at the resulting graph in Fig. 1.1. The X-axis is in logarithmic scale to be able to see the impact of shifts with different order of magnitude. You can see in the center of the graph that the error between the difference ratio described above and the exact value of the derivative appears for shift larger than 10^{-4} but disappears for shifts smaller than 10^{-5}. For the larger shifts, the higher order terms of the function disturb the computation of the exact differentiation. In this case the higher orders have a significant second order part and the forward and backward finite difference errors have similar sizes but with different signs.

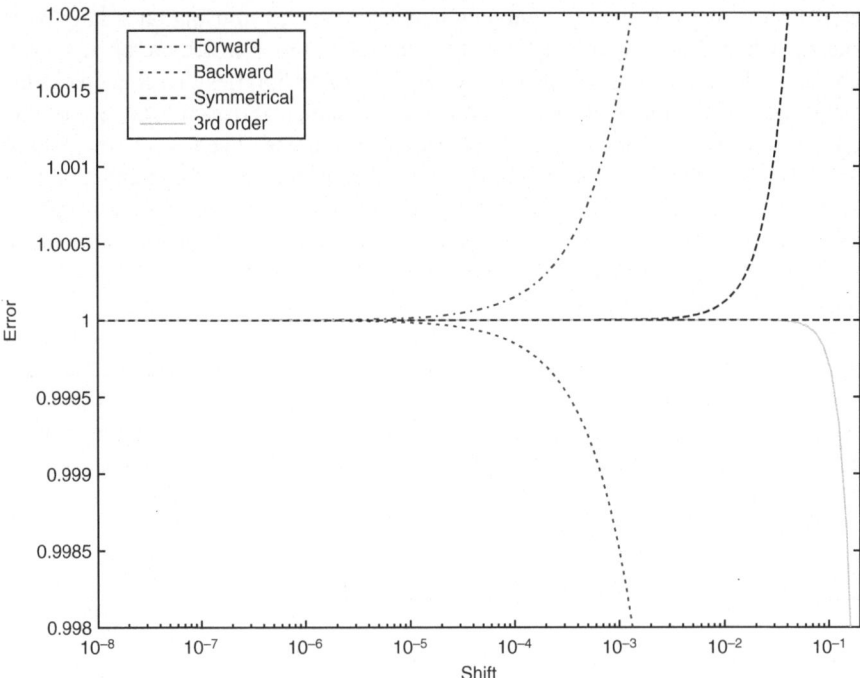

Fig. 1.1 Comparison of the convergence for different finite difference approximation of the first order derivative

To deal with higher order term impact, higher order schemes with better convergence properties can be used. A scheme of order 2 is given by the *symmetrical* (or *centered* or *two sided*) finite difference approximation

$$Df(a) = \frac{f(a+\epsilon)-f(a-\epsilon)}{2\epsilon} + O(\epsilon^2).$$

As can be seen from the same figure, with this approach the convergence happen for larger shifts, and for the example function it is for shifts below 10^{-2}. The computational cost of this approach is one computation for the initial value and two computations for the values with shifted inputs. The total complexity is $1+2*n$ for an n-input function.

One can go one step further and use a scheme of *order four* given by

$$Df(a) = \frac{-f(a+2\epsilon) + 8f(a+\epsilon) - 8f(a-\epsilon) + f(a-2\epsilon)}{12\epsilon} + O(\epsilon^4).$$

As can be seen from the same figure, with this approach the convergence happen for even larger shifts, and for the example function it is for shifts below 10^{-1}. The computational cost of this approach is one computation time for the initial value and four computations for the values with shifted inputs. The total complexity is $1+4*n$ for a n-input function.

We have proposed above four different approaches to the computation of the derivatives of a function using finite difference. As you can guess from the title of the book, none of them is an approach we recommend through this book. Nevertheless it is useful to review them quickly in term of convergence.

Note also that in several places, the one sided finite difference approximation is confused with the real derivative. This is in particular the case in the recent policy framework on Fundamental Review of the Trading book by the Basel Committee on Banking Supervision (2014). Hopefully the Committee will correct this in the final version of the regulation following the advices in my answer to their request for comments. There is no reason to force the industry to use sub-optimal methods when better approaches are available. Central planning and regulators hampering improvements (and the sales of this book!) is unfortunately not unheard off.

The most talked about convergence in place describing the convergence of the approximation methods is below which level of shift ϵ will the approximation be good enough. This is the question we have analyzed above. The function used in the graph is $\exp(x + x^2)$. There is nothing special about this function. The exponential is often used in curve description and $\exp(x^2)$ is related to the normal distribution density which is the base of the diffusion models; the function used is not alien to finance.

From the graph, it is clear that the higher order approximations provide a better approximation of the derivative. A quick answer could be that the first two approximations are good below 10^{-4}, the symmetric one below 10^{-2} and the fourth order one below 10^{-1}. This represents a gain of two orders of magnitude in the shift for the symmetrical approach and three orders for the higher order one. Those gains in precision have to be balanced with the computation cost. The first two approaches have a cost of one computation for each derivative, the symmetrical one has a cost of two and the fourth order one has a cost of four. This looks like a standard trade-off between quality and cost.

But the view of the question at this stage is very one-sided. We have looked at the "right half" of the graph: how large can the shift be and still provides a decent answer? We should look also at the left side of it: how small can the shift be and still provides a decent answer?

The formulas above and their order of convergence are true in theory, i.e. true for infinitely precise arithmetic. In practice, in computers, the precision is limited. Depending of the machine, language, and other details the precision will differ but will be finite in all cases. The limit process involved in the derivative computation by differential ratio is a limit of the style "0/0"; both the numerator and the denominator converge to 0. At some stage, for very small shifts, the figures computed will be indistinguishable by the computer from 0 in its finite precision. The only number left will be the quasi random last bits of the finite precision binary information. The computed ratio becomes of the style "random/random." No meaningful information is left in the process.

This can be seen in Fig. 1.2. The left side of the graph is the computed derivative approximation using the same approximation schemes and letting the ϵ becoming very small. At around 10^{-11} some noise appears and at 10^{-13}, there is more noise than information. This is the case for all the finite difference schemes, even for the higher order ones. The levels creating the problems are the same for all approaches; there is no advantage for the higher order schemes.

Looking at the graph, one may think that this is not a problem, that there is a very wide range of shift values for which all the schemes propose a good enough approximation, just pick one of them. In the above case any shift between 10^{-10} and 10^{-6} is good enough for all practical purposes.

This is certainly true here, with the hindsight of the graph and having run the process hundreds of times for different values. But the answer to the acceptable range question will change for each function; depending of the higher order derivative size, the right size of the range may change dramatically. Close to the money and close to expiry, options have a very high gamma and the above picture on the right would be quite different for them.

The function we used to draw the graph is implemented as an explicit function depending on very easy primitives (exponential and square). In a lot of cases the function we are interested in in finance will be implemented through some more complex numerical schemes. It can involve solving a root-finding problem (like curve calibration or swaption pricing in Hull-White model) or an heavy numerical scheme like numerical integration, PDE solving, trees or Monte Carlo. In that case the boundary between acceptable and unacceptable on the left part of the graph could be a lot more to the right. There is no guarantee that there is any space between the left side of the acceptable range and the right side of it. Even if there is such a range, there is no method a priori to guess where it is. If for each computation,

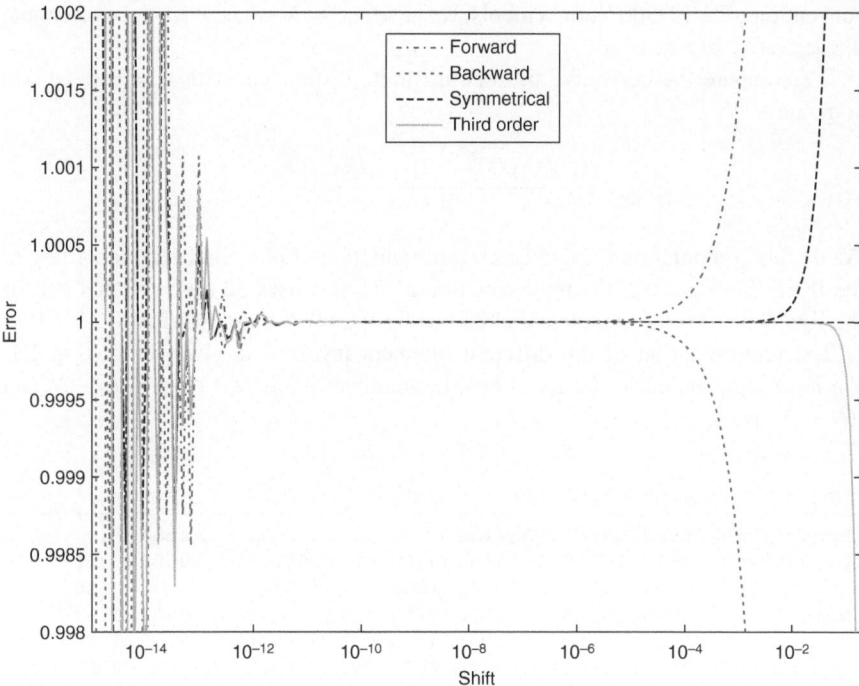

Fig. 1.2 Comparison of the convergence for different finite difference approximations of the first order derivative. Including the "very small shifts" part of the graph

one has to run tens of similar computation, which different shifts, to assess if the initial computation is good enough, any derivative computation would be even more CPU intensive. The problem of finite difference derivatives with trees is described in Pelsser and Vorst (1994) and Henrard (2006). For that numerical technique the short answer is that there is no value of the shift for which a simple bump-and-recompute provides a good result.

We now give an example of what is happening at the computer internal representation level. The goal of this example is to show in the simplest example possible that the behavior described above is not related to the particular function we have used; even the simplest of the functions will display the same problems[2]. We take the example of a bill with a face value of one million currency units, denoted N and three days to maturity. The bill is quoted on a discounted basis, i.e. the value of the bill is $N \times (1 - \delta R)$ with δ the accrual factor and R the rate. We want to compute the sensitivity of the value with respect to a movement of the rate by one basis point. To simplify the matter further, suppose that the current rate is 0 and the day count

---

[2] The example below is inspired by (Naumann 2012, Section 1.3)

convention is "ACT/300"[3]. In symbols, we have $f(x) = N-x$ and we want to compute the derivative of f at $x = 0$.

We compute the derivative by forward finite difference, with $\epsilon = 0.1$, i.e. we compute

$$\frac{(1,000,000 - 0.1) - 1,000,000}{0.1}.$$

We do this computation using float representation of real numbers. According to the IEEE 754 standard, the representation of float uses 32 bits[4] , with 1 bit for the sign, 8 bits for the exponent and 23 bits for the fraction or mantissa.

The representation of the different numbers involved is given in Listing 1.1. We have represented in the five lines the numbers N, ϵ, $N - \epsilon$, $(N - \epsilon) - N$ and $((N - \epsilon) - N)/\epsilon$.

Listing 1.1 Internal representation of float numbers

Theoretical	S-EXPONANT-MANTISSA..............	Actual
1000000.000	0-10010010-11101000010010000000000	1000000.000
0.100	0-01111011-10011001100110011001101	0.100
999999.900	0-10010010-11101000010001111111110	999999.875
-0.100	1-01111100-00000000000000000000000	-0.125
1.000	0-01111111-01000000000000000000000	1.250

Already for the extremely simple case the finite difference in finite precision arithmetic gives a result which is far off with respect to the actual numbers. In this example the base function f requires only one operation, one of the simplest operation for a computer, a subtraction. For a reasonable shift of 0.1, the computed number is 1.25, this is 25% off with respect to the correct number which is 1.

Nevertheless, in a lot of places, the finite difference approach seems to be the state of the art in finance. A lot of derivatives, sensitivities or greeks are computed, in a lot of financial institutions, using a finite difference scheme as described above. Usually without really checking if the approximation is acceptable or not. Maybe twisting the shift to be larger or smaller here and there to get more meaningful numbers.

As you know, or at least as you hoped by buying this book, it is possible to do a lot better and a lot faster. Practical recipes on how to do it in quantitative finance are the *raison d'être* of the book you have in hands.

The comparison in term of performance between algorithmic differentiation and other methods will often be against one-side finite difference (forward or backward). The results will be theoretical for the finite difference. On top of the computation

---

[3] This is an invented convention which is used to simplify the example.
[4] See for example http://en.wikipedia.org/wiki/Single-precision_ floating-point_format.

of the function itself with shifted input, the finite difference requires some data manipulation, in particular it includes copying the input vector to shift one component. For very simple functions, this can take a non-negligible amount of time. In most of our comparisons we ignore those extra-costs. Only in the very first case, with results in Table 2.4, do we include the actual implementation numbers.

In this introduction we used a one dimensional function and computed one derivative. In practice, function used in finance have several inputs, their number often ranging to the hundreds. The computation indicated above should be done for each input, in the finite difference approach, each input to the computation should be bumped individually. The computation time growths linearly with the number of inputs.

1.6   What Is Not in This Book: Higher Order

In the book we discuss the computation of first order derivatives. One may say that in theory, the second order derivative is simply the first order derivative applied to the function implementing the first oder derivative and nothing more should be said. The practice is unfortunately quite different.

Let's start with the dimensions. We said that when we have a several-inputs-and-one-output function, the most efficient approach is the adjoint mode and when the function has one input and several outputs, the most efficient approach is the forward mode. In general in finance, we have a several-inputs-and-one-output function. The derivative function of that function is a several-inputs-and-several-output function; one output for each input of the original function. For those n-input and n-output function, the computation time improvement of AD with respect to finite difference is reduce to nothing and is even not an advantage anymore. The finite difference computation time will be roughly $n + 1$ times, while the *forward AD* method will be around $1.5 \times n$ times and the *adjoint AD* method around $3 \times n$. So the speed advantage is lost for the second order derivatives.

Then there is the fact that the first order derivative is "additive" in the sense that you can compute several part of it separately and add them later. In finance, you can for example compute the "delta" of options in one function and the "vega" in another. This is not the most efficient from a computational point of view but represent the usual business decomposition of risk in several parts. For the second order, this is not true anymore; you cannot compute part of the second order derivatives in one place and another part in a second place, you would be missing the crosses. In financial terms you can compute the cross-gamma somewhere and the cross-volga somewhere else, but you are them missing the crosses between delta and vega.

In Chapter 4, we will discuss AD by operator overloading. The results are obtained by adding information to the doubles used for the computations. This means that the first order derivative code does not exists anywhere, but only the procedures are mocked in memory to achieve the result. The situation is not the same between the original code and the one used for the first order derivatives. We cannot apply the same technique a second time.

For those reason, the technique I use the most often in practice is to compute first order derivatives using AAD and to compute the second order derivatives, when needed, using finite difference on the first order derivatives. We still gain a lot of time and precision with respect to a pure finite difference. On the other side, developing a full second order machinery without a real gain, certainly from a computation time point of view, would require a lot of development time.

1.7   What Is Not In This Book: Code

In the book we propose some (pseudo-code) extract to give practical details of some implementations. The listings do not contain the full code, only enough information to understand what the code does. Nevertheless the full code of a lot of examples is available under an open source license. There is not even the need to type the code from the example, you can download it and run it directly. The code is available from GitHub at the address

```
https://github.com/marc-henrard/
algorithmic-differentiation-book
```

The code proposed also contains an implementation of automatic Algorithmic Differentiation using forward mode and adjoint mode. The implementation is probably not production ready in the sense it is not really an operator overloading implementation but an "operators replaced by methods" implementation. Also the implementation does not propose an optimization of memory allocation; the memory requirement for numerically intensive computation can be a problem in traditional automatic AD implementation. The implementation proposed is nevertheless certainly good enough to understand the advantages and drawbacks of the methods and play with different implementation solutions. The performances results of my automatic AD implementation are in line with advertised results from third party close source implementations.

In some places in the book, for more complex cases, in particular curve sensitivity, curve calibration and model calibration, I refer to OpenGamma's *Strata* library to which I'm one of the contributors. The Strata library is also available open source under the same Apache 2.0 license[5]. My goal in the AD library is to show in action the different implementation techniques, not to develop a full quantitative finance library. The Strata library has a different goal, which is to develop a full-size quantitative finance library using one version of AD as one of its underlying techniques. The examples based on Strata are also available under an open source license at

```
https://github.com/marc-henrard/analysis
```

That repository contains examples related to this book and also other analysis and in particular code related to posts on my blog.

---

[5] It is available on github at the address `https://github.com/OpenGamma/Strata`.

Bibliography

Basel Committee on Banking Supervision. (2014). Fundamental review of the trading book: outstanding issues. Consultative document, Basel Committee on Banking Supervision.

Capriotti, L. (2011). Fast greeks by algorithmic differentiation. *The Journal of Computational Finance*, 14(3):3–35.

Capriotti, L. (2015). Likelihood ratio method and algorithmic differentiation: Fast second order greeks. *Algorithmic Finance*, 4, 4:81–87.

Capriotti, L. and Giles, M. (2010). Fast correlation Greeks by adjoint algorithmic differentiation. *Risk*, 23(3):79–83.

Capriotti, L. and Giles, M. (2012). Algorithmic differentiation: Adjoint Greeks made easy. *Risk*, 24(9):96–102.

Capriotti, L., Lee, J., and Peacock, M. (2011). Real time counterparty credit risk management in Monte Carlo. *Risk*, 24:86–90.

Capriotti, L., Jiang, Y., and Macrina, A. (2015). Real-time risk management: An AAD-PDE approach. *Journal of Financial Engineering*, 2(4).

Capriotti, L., Jiang, Y., and Macrina, A. (2016). AAD and least squares Monte Carlo: Fast Bermudan-style options and XVA greeks. Working Paper Series 2842631, SSRN.

Denson, N. and Joshi, M. (2009a). Fast and accurate Greeks for the Libor Market Model. *Journal of Computational Finance*, 14(4):115–140.

Denson, N. and Joshi, M. (2009b). Flaming logs. *Wilmott Journal*, 1:5–6.

Giles, M. and Glasserman, P. (2006). Smoking adjoints: Fast Monte Carlo Greeks. *Risk*, 19:88–92.

Griewank, A. and Walther, A. (2008). *Evaluating Derivatives: Principles and Techniques of Algorithmic Differentiation*, 2nd edn. Philadelphia:SIAM.

Henrard, M. (2006). A semi-explicit approach to Canary swaptions in HJM one-factor model. *Applied Mathematical Finance*, 13(1):1–18. Preprint available at IDEAS: http://ideas.repec.org/p/wpa/wuwpfi/0310008.html.

Homescu, C. (2011). Adjoints and automatic (algorithmic) differentiation in computational finance. Working Paper Series 1828503, SSRN.

Joshi, M. and Yang, C. (2010). Fast gamma computations for CDO tranches. Working Paper Series 1689348, SSRN.

Kaebe, C., Maruhn, J., and Sachs, E. (2009). Adjoint based Monte Carlo calibration of financial market models. *Finance and Stochastics*, 13(3):351–379.

Leclerc, M., Liang, Q., and Schneider, I. (1999). Fast Monte Carlo Bermudan greeks. *Risk*, 12(7):84–88.

Naumann, U. (2012). *The Art of Differentiating Computer Programs. An Introduction to Algorithmic Differentiation*. Philadelphia:SIAM.

Naumann, U. and du Toit, J. (2014). Adjoint algorithmic differentiation tool support for typical numerical patterns in computational finance. Technical report, RWTH Aachen University.

Pelsser, A. and Vorst, T. (1994). The binomial model and the greeks. *The Journal of Derivatives*, Spring, (13):45–49.

Chapter 2
The Principles of Algorithmic Differentiation

Divide et impera (Divide and conquer): the art of composition. – Where we learn that the world is made of assignments – Two modes: move forward or move backward – Branches

2.1 Derivative

The starting point of everything is obviously the definition of *derivative*, the notion we are planning to compute through Algorithmic Differentiation (AD).

Definition 2.1 (Derivative) A function $f : \mathbb{R}^m \rightarrow \mathbb{R}^n; x \mapsto f(x)$ is said to be *differentiable* at a point $x_0 \in \mathbb{R}^m$ if f is defined on that point and there exist a linear function $Df(x_0) : \mathbb{R}^m \rightarrow \mathbb{R}^n$ such that

$$\lim_{\epsilon \to 0; \epsilon \in \mathbb{R}^m} \frac{f(x_0 + \epsilon) - (f(x_0) + Df(x_0)(\epsilon))}{|\epsilon|} = 0.$$

The linear function $Df(x_0)$ is called the *derivative* of f in x_0.

We will represent element of $\mathbb{R}^n$ by column vectors. The linear functions from $\mathbb{R}^m$ to $\mathbb{R}^n$ will be represented by matrices of dimension m (lines) $\times n$ (columns). Through the text, we will not distinguish between elements of $\mathbb{R}^n$ and column vectors or between linear functions and matrices; the context should be clear enough.

When a function $f : \mathbb{R}^m \rightarrow \mathbb{R}$ is differentiable at a point x_0, we can define its partial derivatives $D_i f(x_0)$ ($1 \leq i \leq m$) as the coefficient of the derivative in the i-th direction. Let e_i be the base vector of $\mathbb{R}^m$ in the i-th dimension, i.e. the vector of $\mathbb{R}^m$ with all components equal to 0 except the i-th one equal to 1. The partial derivative $D_i f(x_0) \in \mathbb{R}$ is defined by $Df(x_0)(e_i)$. In particular we have

$$D_i f(x_0) = \lim_{\epsilon \to 0; \epsilon \in \mathbb{R}} \frac{f(x_0 + \epsilon e_i) - f(x_0)}{\epsilon}. \tag{2.1}$$

By abuse of language, like for matrices, we will use the notation $Df(x_0)$ indifferently for the linear function and the $\mathbb{R}^m$ vector with components $(D_i f(x_0))_{1 \leq i \leq m}$. The context will, in general, be enough to clarify which representation we use.

© The Author(s) 2017
M. Henrard, *Algorithmic Differentiation in Finance Explained*,
Financial Engineering Explained, DOI 10.1007/978-3-319-53979-9_2

The computation of the (partial) derivatives through one-sided finite difference quotient, as described in Section 1.5, is approximating the limit in Eq. 2.1, by the value of the ratio at a fixed value ϵ.

2.2   Composition

The main theoretical concept underlying Algorithmic Differentiation is the notion of *composition*. A long task can be divided in shorter subtasks combined one behind the other. This idea of combining one thing after another is described mathematically by the concept of composition.

Definition 2.2 (Composition)   Let $g : \mathbb{R}^m \to \mathbb{R}^n; x \mapsto g(x)$ and $f : \mathbb{R}^n \to \mathbb{R}^p; y \mapsto f(y)$ be two functions. The function called f *compounded with* g or f *after* g and denoted $f \circ g$, is the function $\mathbb{R}^m \to \mathbb{R}^p$ with

$$(f \circ g)(x) = f(g(x)).$$

This is the same idea as the division of labor. Different tasks are performed by different workers or different functions. Combined specialized functions – or workers, or methods, or micro-services – provide efficiency gains.

One of the main tasks of quantitative finance libraries is to perform the computation of single output functions applied to financial instruments and market data. Throughout the book we use the present value case in the descriptions as it is the most common case in practice, but it is also valid for par rate, currency exposure, price, etc. The use of present value is for convenience in the description, but it does not change the technical part of the analysis.

We suppose that the task of computing this single output method has been implemented using a long list of specialized functions or methods like f and g above. The functions can, and, in practice, will, have several inputs. How to write the algorithm to compute the present value is not the problem that we propose to solve through this book; the reader is expected to have solved that problem already and we suppose that he knows how to implement the value part.

The next task, and the center of this book, is the computation of the sensitivities of those present values to the multiple inputs. In finance those sensitivities are often called the *greeks*. The greeks can be a simple one dimensional number, like the delta of an option to its unique underlying, or multi-dimensional results like the bucketed sensitivities to market quotes in interest rate instruments. The latter being more demanding in computational time and the center of interest here.

It means that the task of this book is to obtain for each function, not only the computation of the value it is supposed to compute, but also of its derivatives with respect to the inputs. If the Black function, providing the option price from the forward, volatility, strike and time to expiry is implemented, we also want to have the derivatives of the price with respect to those four inputs implemented. In the greeks jargon, they are called delta, vega, adjoint delta and theta.

The principles to achieve those implementations are described in this chapter. The building block of this approach is the mathematical result below[1].

Theorem 2.1 (Chain rule) Let $g : \mathbb{R}^m \to \mathbb{R}^n; x \mapsto g(x)$ and $f : \mathbb{R}^n \to \mathbb{R}^p; y \mapsto f(y)$ be two functions. If g is differentiable in x_0 and f is differentiable in $g(x_0)$, then the composition is differentiable in x_0. The derivative of the composition is given by

$$D(f \circ g)(x_0) = Df(g(x_0)) \cdot Dg(x_0). \tag{2.2}$$

In Formula (2.2), the operation $\cdot$ is the composition if we use the linear function representation and is the matrix multiplication if we use the matrix representation.

The differentiation operator transforms a composition of function into the composition of the linear functions representing the derivatives. The composition of linear functions can be computed through the product of the matrices representing the functions.

The result on the derivative of a composition of functions may look benign, with simply a circle on the left hand side of the equality replaced by a small dot on the right hand side. But that simple one liner is the core of AD; understanding that one line and its impact on quantitative finance libraries is the only goal of this book. This one line is a tremendously good news for library's architects!

The summary of the impact is that the differentiation part of the library can be written without changing the architecture put in place for the valuation. The composition of functions is transformed into a composition of linear operators, which can be written as a product of matrices. Any architecture that has been established and is efficient for computing the values by splitting the general task in subtasks can be used without change for the computation of the derivatives. The method achieving the different tasks will have to be complemented by new methods to achieve the differentiation of those tasks, but the general architecture can stay unchanged.

As one good news never comes alone, the second good news is the product part. The composition of function at the value level is converted in a product of matrices at the derivative level. The product of matrices is, at the coefficient levels, a combination of sums and products of real numbers. Those real numbers are represented in computers by `double`. It turns out that processors are very efficient at computing sums and products of doubles. We can thus expect that there will be a very hardware-efficient way to implement the derivative computation procedure deduced from the composition formula.

Now it is time for a little bit of bad news. Up to now we have said that the computation of derivatives can be done without changing the architecture and can probably be done very efficiently. The bad news is that it has to be done; it will not

---

[1] We describe the results without all the rigor that would be used in a mathematical text. In particular, we do not indicate which exact class of functions will satisfy the results. The functions are supposed regular enough to satisfy the different results, without indicating exactly what we mean by regular enough. More precise statement are given in Appendix A.

appear magically. It is not only a figure of speech that "something has to be done" but that to have it working everything has to be done.

When you read the theorem again, you can see that the final result requires the derivatives of each of the composing functions. If you want to implement this technique for your library you have to start from the very bottom and work your way up. There is no way to get the benefits for a couple of time consuming pricing methods at the top of your library and ignore the rest. All the low level pieces on which those complex pricing methods are based need to be implemented in an AD compatible way, with their derivatives. In the quantitative finance case, this often mean that interpolation schemes, discount factors, Black formula, forward curves, and other building block methods have to be looked at in all details before we start looking at numerical integration, option pricing, Monte Carlo or other numerically demanding tasks.

The Algorithmic Differentiation approach is really a *philosophy* based on the division of labor. Each small task is performed by an expert method providing the value and all its derivatives with respect to its inputs. As much as it was emphasized in the previous paragraph that AD is a bottom up approach that requires all the methods to provide the derivatives, as much there is nobody looking under the hood to see how the task is performed. Each subtask can provide the derivatives using its own technique. For some subtask, one may choose to use either a finite difference method as described in the introduction, or an adjoint AD implementation, or a standard AD implementation, or a direct analytical formula. There is no restriction on how each part of the list of tasks is working internally, as long as it provides the required result in an efficient way. An Algorithmic Differentiation philosophy could be implemented with only finite difference at the lowest level but product of derivatives at each level above. There are as many shades of AD as there are AD users. The *Principles of Algorithmic Differentiation* chapter provides to the user the black and the white; it is up to him to get the correct shade of grey that fits his taste and his requirements.

This remark is quite important in term of implementation. If a financial instrument appears only once in the balance sheet of a bank and there is no incentive to improve the computation speed for it, its derivatives can be computed by a simple finite difference approach. This will not impact the general functioning of the library. The derivatives are provided, maybe inefficiently, but they are there and the global process can be achieved. For example when implementing a new interpolation mechanism for a yield curve, one may chose to focus on the results first without looking at the computational efficiency by implementing the derivatives of the interpolation using a generic finite difference implementation. If the results are good from a financial perspective, one may come back to the details of the interpolation derivatives latter. The internal implementation may be changed but it should remain transparent to the developer. Each method has a value implementation and a value and derivatives implementation. Those methods can be refined internally but its availability to the outside world and its signature remain the same.

2.3   Algorithm: Assignment

We now enter into the internal working of the Algorithmic Differentiation principles. We start discussing code writing.

The starting point is the computation of a function $f : \mathbb{R}^{p_a} \to \mathbb{R}$;

$$a \mapsto z = f(a). \tag{2.3}$$

The function inputs are $a = a[0 : p_a - 1]$ of dimension p_a. We use a mixture of Matlab and Java notation for the vectors, with $a[i : j]$ being the vector of dimension $j - i + 1$ and components a_k for $k = i, \ldots, j$.

The vector has to be thought of as the market quotes used to calibrate curves or as the inputs used to price options, like forward, volatility – single value or full surface –, strike, bond price, etc. The output of the function is z of dimension 1. The main application in finance is for the present value, the price, the implied volatility, the spread or the par rate.

As a starting point we suppose that the program that represents the algorithm to compute f is a simple linear program with one assignment for each line of code and nothing else. This approach is called Single Assignment Code (SAC) in Naumann (2012). We will introduce other potential elements in the code later. In particular ifs, loops and equation solving.

As the starting point of the analysis of Algorithmic Differentiation, we suppose that the algorithm for the function itself is implemented. It is not the goal of this book to explain how to implement the algorithms for the functions themselves.

All the intermediary values used in the program will be denoted b (with different indices). The algorithm starts with the inputs a and goes to z through a lot of bs. The new variables are denoted by $b[j]$ with j starting at 1 and going up to p_b. There are p_b intermediary variables (or parameters) b in the program. The number p_b is also the number of lines of code. At each line of code, the new variable $b[j]$, which depends potentially on all the previous variables $b[k]$ with $k < j$ and the initial inputs a, is created.

The simplified version of an algorithm is represented in Table 2.1. Note that we initialize the variables $b[j]$ $(-p_a + 1 < j < 0)$ with the inputs values. This is to unify the notation of the Algorithmic Differentiation implementation later. Each line of code perform the computation of a function g_j. That function can be an elementary arithmetic computation, like $b_1 + b_2$, a mathematical function, like $\exp(b_3)$, or another method already implemented in the library for which the value and derivative versions are available.

Table 2.1 The generic single assignment code for a function computing a single value

Initialization	$[j = -p_a + 1 : 0]$	$b[j]$	$=$	$a[j + p_a - 1]$
Algorithm	$[j = 1 : p_b]$	$b[j]$	$=$	$g_j(b[-p_a + 1 : j - 1])$
Output		z	$=$	$b[p_b]$

2.3.1   Standard Algorithmic Differentiation

Standard Algorithmic Differentiation is also called *Forward Algorithmic Differentiation* or *Tangent Algorithmic Differentiation*.

Our goal is to compute $\partial z/\partial a_i$. We achieve this by computing for each j ($-p_a+1 \le j \le p_b$) the value

$$\dot{b}[j, i] = \frac{\partial}{\partial a[i]} b[j].$$

Note that the derivative is denoted by a dot (˙) on the variable and $\dot{b}[j, i]$ is the derivative of $b[j]$ itself with respect to some other variable, the inputs $a[i]$. The summary of the standard AD algorithm is presented in Table 2.2. The starting point is the easiest part. For $j = -p_a + 1 : 0$, the derivative of $b[j]$ with respect to $a[i]$ is simply the derivative of $a[i]$ with respect to itself – which is 1 – for $j = -i$ and 0 for $j \ne -i$. This is the starting point of a recursive algorithm. The starting part is an identity matrix: $\dot{b}[j, i] = \delta_{i,-j}$.

From there we read the code in the forward order and use the derivative of a composition p_a times for each line of code. Each intermediary variable $b[j]$ uses only the variables that have been populated in the previous lines of code in the implementation of the algorithm. The derivatives $\dot{b}[j, i]$ are given by

$$\dot{b}[j, i] = \frac{\partial}{\partial a[i]} b[j] = \sum_{k=-p_a+1}^{j-1} \frac{\partial}{\partial b[k]} b[j] \cdot \frac{\partial}{\partial a[i]} b[k]$$

$$= \sum_{k=-p_a+1}^{j-1} \frac{\partial}{\partial b[k]} g_j(b[-p_a + 1 : j - 1]) \cdot \dot{b}[k, i].$$

From the recursive approach, the values $\dot{b}[k, i]$ for $k < j$ are already known. The derivatives of the functions g_k can be obtained from the implementation of algorithmic

Table 2.2 The generic code for a function computing a single value and its standard algorithmic differentiation pseudo-code

Initialization	$[j = -p_a + 1 : 0]$	$b[j]$	$= a[j + p_a - 1]$
Algorithm	$[j = 1 : p_b]$	$b[j]$	$= g_j(b[-p_a + 1 : j - 1])$
Output		z	$= b[p_b]$
Initialization	$[j = -p_a + 1 : 0][i = 0 : p_a - 1]$	$\dot{b}[j, i]$	$= \delta_{i,-j}$
Algorithm	$[j = 1 : p_b][i = 0 : p_a - 1]$	$\dot{b}[j, i]$	$= \displaystyle\sum_{k=-p_a+1}^{j-1} \frac{\partial}{\partial b_k} b_j \frac{\partial}{\partial a_i} b_k$
			$= \displaystyle\sum_{k=-p_a+1}^{j-1} \frac{\partial}{\partial b_k} g_j \dot{b}[k, i]$
Output	$[i = 0 : p_a - 1]$	$\dfrac{\partial}{\partial a_i} z$	$= \dot{b}[p_b, i]$

differentiation that we have supposed to have been done for all the functions g_j. Using that approach one can recursively obtain the values $\dot{b}[j, i]$ up to $\dot{b}[p_b, i]$ going through $j = -p_a + 1$ to p_b. The numbers $\dot{b}[p_b, i]$ are equal to the derivatives of $z = b[p_b]$ with respect to a_i ($i = 0, \ldots, p_a - 1$). This concludes the algorithm for the computation of $\partial z/\partial a_i$.

The requirements for such an implementation is that all the intermediary functions g_j have a derivative version. The Algorithmic Differentiation approach is a bottom-up approach: it can be implemented for an algorithm only if all the components "below it," all the components entering into the composition have already been implemented.

There is no way to take the most complex function in a library and try to implement algorithmic differentiation there if it has not been implemented fully for all the functions below. It is not a miraculous method that can be applied in complex situations and ignored in simpler, more elementary parts of a library. You cannot have a solid quantitative finance library if you have not spend enough time and efforts on the foundations.

As a first practical example of the theoretical description above, we implement the algorithm for a very simple function. The function has no special financial meaning. It is a four inputs one output function using elementary analytical functions.

The function (with 4 inputs) is:

$$z = \cos\left(a_0 + \exp(a_1)\right)\left(\sin(a_2) + \cos(a_3)\right) + (a_1)^{\frac{3}{2}} + a_3.$$

The code of one algorithm to compute that simple function is given in Listing 2.1. As indicated in Section 1.7, the full code for the examples is available in an open source format on Github. From now on, we don't repeat the advertisement for the open source code and we simply give the name of the class in which the code can be found. The code corresponding to Listing 2.1 can be found in the class `AdStarter`.

Listing 2.1 Simple function code

```
double f(double[] a) {
    double b1 = a[0] + Math.exp(a[1]);
    double b2 = Math.sin(a[2]) + Math.cos(a[3]);
    double b3 = Math.pow(a[1], 1.5d) + a[3];
    double b4 = Math.cos(b1) * b2 + b3;
    return b4;
}
```

The code for the manual Standard Algorithmic Differentiation version is given in Listing 2.2 and in the method `f_Sad` of the class `AdStarter`. Note that the output of the original function is a double while the output of the AD version is an object containing a double and an array representing the derivatives of the result with respect to the different inputs. The description of the object, called `DoubleDerivatives` can be found in Listing 2.3.

Listing 2.2 Standard Algorithmic Differentiation version of the starter function

```
static public DoubleDerivatives f_Sad(double[] a) {
    // Forward sweep - function
    double b1 = a[0] + Math.exp(a[1]);
    double b2 = Math.sin(a[2]) + Math.cos(a[3]);
    double b3 = Math.pow(a[1], 1.5d) + a[3];
    double b4 = Math.cos(b1) * b2 + b3;
    // Forward sweep - derivatives
    int nbA = a.length;
    double[] b1Dot = new double[nbA];
    b1Dot[0] = 1.0;
    b1Dot[1] = Math.exp(a[1]);
    double[] b2Dot = new double[nbA];
    b2Dot[2] = Math.cos(a[2]);
    b2Dot[3] = - Math.sin(a[3]);
    double[] b3Dot = new double[nbA];
    b3Dot[1] = 1.5d * Math.sqrt(a[1]);
    b3Dot[3] = 1.0d;
    double[] b4Dot = new double[nbA];
    for(int loopa = 0; loopa < nbA ; loopa++) {
        b4Dot[loopa] = b2 * - Math.sin(b1) * b1Dot[loopa] +
            Math.cos(b1) * b2Dot[loopa] + 1.0d * b3Dot[loopa];
    }
    return new DoubleDerivatives(b4, b4Dot);
}
```

Listing 2.3 Object containing a double and its derivatives with respect to the inputs

```
public class DoubleDerivatives {
    private final double value;
    private final double[] derivatives;

    public DoubleDerivatives(double value, double[] derivatives) {
        this.value = value;
        this.derivatives = derivatives;
    }
    ...
}
```

The code itself does not require lengthy explanations. For each variable bx, a derivative vector is created with the dimension of the inputs, which is 4 and denoted nbA in the code. When created, the vector is filled with 0, which is the default in Java. Then we look at each of the next lines and see how they depend on the inputs. The first three lines depend only on the inputs and the derivatives of standard functions are used. The fourth variable depends of the previous variables.

Its derivatives depends on the four inputs; they are computed in a loop on the dimension of the inputs.

The performance of the different implementations in term of computation time is discussed at the end of the section and summarized in Table 2.4. The performance is analyzed in the test class `AdStarterAnalysis`.

2.3.2 Adjoint Algorithmic Differentiation

Adjoint Algorithmic Differentiation is also called *Reverse Algorithmic Differentiation*.

Our goal is to compute $\partial z/\partial a_i$. We achieve this by computing for each intermediary variable j $(-p_a + 1 \le j \le p_b)$ the value

$$\bar{b}[j] = \frac{\partial}{\partial b[j]} z.$$

Note that the derivative is denoted by a "bar" ($\bar{\ }$) on the variable and $\bar{b}[j]$ is the derivative of the output with respect to $b[j]$. This is a non-standard notation in the sense that the variable in the name is not the variable that is differentiated but the variable with respect to which the derivative is taken. It is important to switch perception between the forward and reverse approach. What is fixed in the reverse approach is the output; we always compute the derivative of the same variable, the output; it is not required to repeat that information at each step.

The summary of the algorithm is presented in Table 2.3. The starting point of the algorithm is the easy part. For $j = p_b$, the derivative of z with respect to b_j is simply the derivative of z with respect to itself, which is 1. This is the starting point of a recursive algorithm.

Table 2.3 The generic code for a function computing a single value and its adjoint algorithmic differentiation code

Init	$[j = -p_a + 1 : 0]$	$b[j]$	$= a[j + p_a - 1]$
Algorithm	$[j = 1 : p_b]$	$b[j]$	$= g_j(b[-p_a + 1 : j - 1])$
Value		z	$= b[p_b]$
Value		$\bar{z}$	$= 1.0$
Value		$\bar{b}[p_b]$	$= 1.0$
Algorithm	$[j = p_b - 1 : -1 : -p_a]$	$\bar{b}[j]$	$= \displaystyle\sum_{k=j+1}^{p_b} \frac{\partial}{\partial b_k} z \frac{\partial}{\partial b_j} b_k$
			$= \displaystyle\sum_{k=j+1}^{p_b} \bar{b}[k] \frac{\partial}{\partial b_j} g_k$
Init	$[i = 0 : p_a - 1]$	$\dfrac{\partial}{\partial a_i} z$	$= \bar{b}[i - p_a + 1]$

From there we read the code in the reverse order and uses the derivative of a composition. Each intermediary variable $b[j]$ is used only in the lines of code that follow in the computation. The derivative $\bar{b}[j]$ is given by

$$\bar{b}[j] = \frac{\partial}{\partial b_j}z = \sum_{k=j+1}^{p_b}\frac{\partial}{\partial b_k}z \cdot \frac{\partial}{\partial b_j}b_k = \sum_{k=j+1}^{p_b}\bar{b}[k] \cdot \frac{\partial}{\partial b_j}g_j(b[-p_a+1:k-1]). \quad (2.4)$$

From the recursive approach, the values $\bar{b}[k]$ for $k > j$ are already known. The derivatives of the functions g_k can be obtained from the implementation of algorithmic differentiation for the building blocks functions g_k or as known mathematical functions. Using that approach one can recursively obtain the values $\bar{b}[j]$ down to $\bar{b}[l]$ for $l = -p_a + 1, \ldots, 0$. Those numbers are equal to the derivatives of z with respect to the inputs a_i ($i = 0, \ldots, p_a - 1$). This concludes the algorithm for the adjoint mode of AD.

The version of adjoint Algorithmic Differentiation for the simple function described above is displayed in Listing 2.4 and implemented in the method f_Aad of the class AdStarter.

Listing 2.4 Adjoint Algorithmic differentiation version of the function

```
static public DoubleDerivatives f_Aad(double[] a) {
  // Forward sweep - function
  double b1 = a[0] + Math.exp(a[1]);
  double b2 = Math.sin(a[2]) + Math.cos(a[3]);
  double b3 = Math.pow(a[1], 1.5d) + a[3];
  double b4 = Math.cos(b1) * b2 + b3;
  // Backward sweep - derivatives
  double[] aBar = new double[a.length];
  double b4Bar = 1.0;
  double b3Bar = 1.0d * b4Bar;
  double b2Bar = Math.cos(b1) * b4Bar + 0.0 * b3Bar;
  double b1Bar = b2 * - Math.sin(b1) * b4Bar
    + 0.0 * b3Bar + 0.0 * b2Bar;
  aBar[3] = 1.0 * b3Bar - Math.sin(a[3]) * b2Bar + 0.0 * b1Bar;
  aBar[2] = Math.cos(a[2]) * b2Bar;
  aBar[1] = 1.5d * Math.sqrt(a[1]) * b3Bar
    + Math.exp(a[1]) * b1Bar;
  aBar[0] = 1.0 * b1Bar;
  return new DoubleDerivatives(b4, aBar);
}
```

The dimension of the returned vector $\bar{a}$ is the same as the one of a. The function returns the final value b4 and returns its derivatives in aBar. The value 0.0 * bxBar should be added in a lot of places for the aBar computations. Here we ignore those terms that appear in the theoretical algorithm but do not add any value in practice.

Table 2.4 Computation time for the example function and four derivatives

Repetitions	Computation	Time	Ratio
100,000	Function	110	1.00
100,000	Function + FD forward	575	5.23
100,000	Function + FD backward	575	5.23
100,000	Function + FD Symmetrical	915	8.32
100,000	Function + FD 4th order	1830	16.64
100,000	Function + SAD	340	3.09
100,000	Function + SAD Optimized	195	1.77
100,000	Function + AAD	220	2.00
100,000	Function + AAD Optimized	165	1.50

Figures for time in milliseconds. Times measured on the author's laptop. Each repetition is with five different data sets so that there is actually 500,000 repetitions.

Note that the mathematical derivative of exp, sin and cos are known and already implemented. In a full library implementation, one is not supposed to know what the derivatives of the other methods are; all methods should have a relevant derivative version available.

What is the performance of those different implementations? The estimation of the computation times are provided in Table 2.4. First we compute the derivatives using different finite difference schemes. This allows us to check that our implementation of the AD versions are correct – unit tests are important in a library – or at least that they are correct within a certain error margin due to the systematic error from the finite difference approach. It allows us also to compare the performance on a practical case and see if our above theoretical discussion related to the finite difference performance is realistic. The performance is analyzed in the test `derivativesPerformance` of the class `AdStarterAnalysis`. Before the run used to produce the table, the Java JVM is *warmed-up*, allowing it to find the hotspots and JIT these. This will be the case for all the performance tests discussed in the book.

For the finite difference versions, the performance is in line with our description. The one-sided schemes, the forward and the backward finite difference, take slightly more than $1 + n = 5$ times the CPU time used for the value computation. For the two-sided scheme the computation time is slightly more than $2 * n = 8$ and for the fourth order scheme, the computation time is slightly more than $4 * n = 16$. As can be seen from the actual numbers, the finite difference take a little bit more time than the theoretical description. This is the time required to manipulate the data before starting the repricing; in particular the input data vector needs to be copied and shifted by the finite difference shift.

If we look at the different Algorithmic Differentiation versions, we see that the results are in line with promises. The direct adjoint implementation takes roughly two times the value computation time for computation of the value and the four derivatives. The standard implementation is a little bit slower with a ratio above three. This result for standard algorithmic differentiation is not typical; usually the ratio is closer to the number of inputs plus 1 (5 in our case). This improved performance is due to the very simplistic nature of our example. For advanced case, in particular the Black option price function discussed in the next chapter, will show that this is not a general result.

We have introduced two new version of the AD implementation. In one of them, the adjoint version code has been "optimized" in a very simple way. The code reuses as much as possible the numbers already computed. In particular, the `Math.exp(a[1])` is used twice in the first code. Computing it once and storing the value for the second usage saves some computation time. Similarly, storing `Math.cos(b1)` helps. We also store `Math.pow(a[1], 1.5d)` and notice that `Math.sqrt(a[1])` is equal to `Math.pow(a[1], 1.5d) / a[1]`. The division being faster than the square root, we gain a little bit more time. In total, with those small improvements, the computation time ratio is now around 1.50. It means it is more efficient to compute *the value and all four derivatives* by AD than to compute *only one derivative* by finite difference. Even if the code is used to extract only one derivative, the adjoint AD version is more efficient that one-sided finite difference, not to mention more precise. This is a feature we will notice repeatedly; you don't need to require all derivatives of a function to take advantage of AD. In a lot of cases, if you need the derivatives to only one or two input, it is more efficient to compute all derivatives – maybe hundreds of them – by AD and keep only the one required than to compute only the one you need by finite difference.

Obviously this example can be seen a "cheating," there is no full library involved, no call to lower level functions with their own AD implementation. In the next chapter, we will introduce more complex situation in financial applications.

2.4   Algorithm: Branches

Algorithmic Differentiation deals with derivatives. It is expected that the function to which the mechanism is applied is itself differentiable. If not the result is undefined. This is the occasion to introduce a warning: *Algorithmic Differentiation creates the derivative of the code executed, not a derivative of the function.* It is important to remember that in the execution, one set of instructions is executed. It is the derivative of that set of instructions that is naturally obtained by AD, not the derivative of the abstract function that is approximated by the code. To parody the saying about *"winners writing history"* one could say that the *"the winning branch is writing the code"* or anticipating on the automatic Algorithmic Differentiation of Chapter 4, *"the winning branch is writing the tape."*

One extreme case would be the implementation of the logarithm function with a huge lookup table (remember those logarithm tables?). It means that in all cases the result returned will be one of the constant of the table. What is the derivative of a constant? It is 0. Such an implementation of the function with a un-careful use of AD would lead to the practical result that the derivative of the logarithm is always 0!

One case where this division of a function algorithm into different sub-algorithms executed under different circumstances appears explicitly, is the presence of

Table 2.5 The generic code for a function computing a single value with an *if* statement

Initialization	$[j = -n + 1 : 0]$	$b[j]$	$=$	$a[j + n - 1]$
Algorithm	$[j = 1 : m]$	$b[j]$	$=$	$g_j(b[-n + 1 : j - 1])$
				$\text{if } f(b[-n + 1 : m])\{$
	$[j = m + 1 : p_b]$	$b[j]$	$=$	$g_j(b[-n + 1 : j - 1])$
				$\} \text{ else } \{$
	$[j = m + 1 : \tilde{p}_b]$	$b[j]$	$=$	$\tilde{g}_j(b[-n + 1 : j - 1])\}$
Value		z	$=$	$b[p_b] \quad \text{or} \quad b[\tilde{p}_b]$

Table 2.6 The generic code for a function computing a single value and its algorithmic differentiation code

Initialization	$[j = -n + 1 : 0]$	$b[j]$	$=$	$a[j + n - 1]$
Algorithm	$[j = 1 : m]$	$b[j]$	$=$	$g_j(b[-n + 1 : j - 1])$
				$\text{if } f(b[-n + 1 : m])\{$
	$[j = m + 1 : p_b]$	$b[j]$	$=$	$g_j(b[-n + 1 : j - 1])$
				$\} \text{ else } \{$
	$[j = m + 1 : \tilde{p}_b]$	$b[j]$	$=$	$\tilde{g}_j(b[-n + 1 : j - 1])\}$
Value		z	$=$	$b[p_b] \quad \text{or} \quad b[\tilde{p}_b]$
Value		$\bar{z}$	$=$	1.0
Value	$\bar{b}[p_b] \quad \text{or} \quad \bar{b}[\tilde{p}_b]$		$=$	1.0
				$\text{if } f(b[-n + 1 : m])\{$
Algorithm	$[j = p_b - 1 : -1 : -p_a]$	$\bar{b}[j]$	$=$	$\sum_{k=j+1}^{p_b} \bar{b}[k] \dfrac{\partial}{\partial b_j} g_k$
				$\} \text{ else } \{$
	$[j = \tilde{p}_b - 1 : -1 : -p_a]$	$\bar{b}[j]$	$=$	$\sum_{k=j+1}^{\tilde{p}_b} \bar{b}[k] \dfrac{\partial}{\partial b_j} g_k\}$
Init	$[i = 0 : p_a]$	$\dfrac{\partial}{\partial a_i} z$	$=$	$\bar{b}[i - p_a + 1]$

branches or *ifs* in the code. Suppose that the code is now something like the one described in Table 2.5 with f a function of $n + j$ double with boolean value.

The resulting AAD code is given in Table 2.6. The branching on the main algorithm is repeated in the backward sweep.

At this stage, one has to be careful with the result's meaning. Suppose that the boolean function denote f in the table is of the type $b[m] > 0$. This will lead to non-differentiable function at the point where $b[m] = 0$ in most of the cases. To be correct, if we are in the case $b[m] = 0$, both branches should be computed and the results of the value itself and its derivatives should be compared. Only if both the values and the derivatives of the different branches are equal, should a derivative result be returned. In practice this is not done. The code is executed, going through one and only one of the branches and the result is returned. It is expected that it is the responsibility of the algorithm writer and not the code writer to verify that this is the case. The actual differentiability of the function resulting from the algorithm is not checked at the code level, it is assume to take place before, in the design phase.

There are cases where the branching, even with only one very simple branch, can be nasty. Take the case where the condition is of the type $b[m] = 0$ and the first branch is something like $b[m+1] = g_m(b[-n+1 : m]) = 0$. This is not an exceptional case, and we will encounter it in our SABR example in the next chapter. If you apply the AD algorithm directly, you get

$$\bar{b}[m-1] = \bar{b}[m]\frac{\partial}{\partial b_{m-1}}g_m = \bar{b}[m]0 = 0$$

$$(j < m-1) \quad \bar{b}[j] = \sum_{k=j+1}^{p_b} \bar{b}[k]\frac{\partial}{\partial b_j}g_k = \sum_{k=j+1}^{p_b} 0\frac{\partial}{\partial b_j}g_k = 0.$$

From that very simple branch, where the value is a hardcoded constant when the branching condition is met for one specific value, we have contaminated the rest of the algorithm. All the derivatives computed from that branch will be 0.

When you look at it from a theoretical point of view, this is the correct result. We act like if the branches did not exist; if we take one branch, only the code in that branch exists. The final result is a constant. The derivative of a constant is always zero, whatever the inputs are. The AD implementation for that branch is a complex mechanism to produce outputs with value 0. It is very difficult to escape from that trap in an automatic way. Except from the algorithm designer knowing the problem and creating a local solution, we don't know a way to solve this. In this case, the local solution would be something like

$$b[m + 1] = c \times b[m]$$

with c the derivative of the output with respect to the input $b[m]$ computed by the algorithm designer outside of the AD framework. In that case, the output will be the same: $b[m + 1] = 0$, but the derivative will be different, it will be

$$\bar{b}[m] = c \times \bar{b}[m + 1] = c.$$

2.5   Algorithm: Loops

Loops, through *for* or *while*, are almost a non-event for Algorithmic Differentiation. If the code in the loop is composed of single assignments, one can unwind the loop to create a longer list of assignments. The assignments may not be written separately from a code perspective, but they are still executed consecutively as if they were different lines of the program.

As the adjoint mode of AD goes through the code in reverse order, the loops have to be also read in the reverse order. A loop that reads as **for** (**int** i = 0; i < N; i++) in the function computation will probably read as **for** (**int** i = N-1; i >= 0; i--) in the reverse AD code.

Moreover, one has to be careful with the variables that change value through the loop. The safest approach is often to change from a single variable that changes value through the loop to an array, with each element assigned only once. This is more in line with the Single Assignment Code presentation used in this chapter.

Bibliography

Naumann, U. (2012). *The Art of Differentiating Computer Programs. An Introduction to Algorithmic Differentiation*. Philadelphia:SIAM.

Chapter 3
Application to Finance

Knowledge of finance and mathematics helps for programming – A single point is not enough to know about derivatives – Function spaces are of infinite dimension – Black magic with Black formula.

It seems there is no easy starting point for reviewing Algorithmic Differentiation in finance. Each function, especially in the simpler one, seems to be an exception with a particular trick more than a general direct application of the methodology.

Instead of trying to find a convoluted way to introduce a simple application that comes directly from the general principles, we introduce a couple of application to finance. For each of them the relevant pricing approaches probably appear in the early chapters of quantitative finance books.

We will look at the Black-Scholes formula and discover that knowing finance and mathematics helps also for Algorithmic Differentiation. Next we move to SABR implied volatility formula to remember that the derivative of a function at a given point does not depend on the point alone but also on its neighborhood. Further, we pass by cash flow discounting to be reminded that function spaces are of infinite dimension which are not easily represented in computers and finish with Monte Carlo simulations that are maybe numerically complex but look straightforward from an AD implementation perspective.

3.1 Black Formula

We start with the Black and Scholes formula, or more exactly the version in term of forward price or rate often referred to as Black (1976). We could have started with the Bachelier (1900) formula, if we were to respect the historical order, but from a popularity perspective, Black formula seems a better starting point. The formula can be summarized in a couple of lines. The inputs are the underlying's forward price or rate F, the (implied) volatility σ, the numeraire P, the strike price or rate K and the time to expiry θ. The formula is different for calls and puts. In the formulas, we use $\omega = 1$ for a call and $\omega = -1$ for a put. The formula is based on the quantities

$$d_\pm = \frac{\ln\left(\dfrac{F}{K}\right) \pm \dfrac{1}{2}\sigma^2\theta}{\sigma\sqrt{\theta}}.$$

© The Author(s) 2017
M. Henrard, *Algorithmic Differentiation in Finance Explained*,
Financial Engineering Explained, DOI 10.1007/978-3-319-53979-9_3

The value d_+ is the boundary between exercise and non-exercise regions in the underlying random normal variable – after change of measure to the numeraire measure. The present value is given by

$$\text{Black}(F, \sigma, P, K, \theta) = \omega P\,(FN(\omega d_+) - KN(\omega d_-)) \tag{3.1}$$

where N is the cumulative standard normal distribution function.

The implementation of such formula is very simple and direct and is represented in Listing 3.1. The implementation code for the Black function can be found in the class `BlackFormula` of the library associated to the book. The implementation does not check for correctness of the inputs – at least the forward, volatility and strike should be positive – as should good production code, but the implementation contains all the numerical ingredients of the formula.

Listing 3.1 Simple implementation of the Black 76 formula

```
double price(double forward, double volatility,
   double numeraire, double strike, double expiry,
   boolean isCall) {
  double omega = isCall ? 1.0d : -1.0d;
  double periodVolatility = volatility * Math.sqrt(expiry);
  double dPlus = Math.log(forward / strike) / periodVolatility
  + 0.5d * periodVolatility;
  double dMinus = dPlus - periodVolatility;
  double nPlus = NORMAL.getCDF(omega * dPlus);
  double nMinus = NORMAL.getCDF(omega * dMinus);
  double price = numeraire * omega
  * (forward * nPlus - strike * nMinus);
  return price;
}
```

Looking quickly at the code, it appears to fit perfectly with our description of SAC code and that nothing more can be said. Let's start with the first part. Yes, it fits perfectly the applicability criterion. The standard and adjoint AD code can be written directly. As the standard mode is not the most efficient in finance, we don't comment on it any more. As a reference, the code is nevertheless available in the code repository in the class mentioned above.

The output of the AD methods are the price and the derivative of the price with respect to the inputs. Like in the previous chapter, the output of the methods are `DoubleDerivatives` objects. The Listing with the description of the object was provided in Listing 2.3.

The adjoint algorithmic differentiation code is provided in Listing 3.2. The new output for the AD adjoint mode provides the derivative with respect to the five inputs – forward, volatility, numeraire, strike, and expiry – in the order they are listed in the method signature.

Listing 3.2   Direct implementation of the Black 76 formula in adjoint mode

```
DoubleDerivatives price_Aad(double forward, double volatility,
    double numeraire, double strike, double expiry,
    boolean isCall) {
// Forward sweep - function - see above
...
// Backward sweep - derivatives
double priceBar = 1.0;
double nMinusBar = numeraire * omega * -strike * priceBar;
double nPlusBar = numeraire * omega * forward * priceBar;
double dMinusBar = NORMAL.getPDF(omega * dMinus)
    * omega * nMinusBar;
double dPlusBar = 1.0d * dMinusBar +
    NORMAL.getPDF(omega * dPlus) * omega * nPlusBar;
double periodVolatilityBar = -1.0d * dMinusBar +
    (-Math.log(forward / strike) /
    (periodVolatility * periodVolatility) + 0.5d) * dPlusBar;
double[] inputBar = new double[5];
inputBar[4] = volatility * 0.5 / Math.sqrt(expiry)
    * periodVolatilityBar;
inputBar[3] = -1.0d / strike / periodVolatility * dPlusBar
    + numeraire * omega * -nMinus * priceBar;
inputBar[2] = omega * (forward * nPlus - strike * nMinus)
    * priceBar;
inputBar[1] = Math.sqrt(expiry) * periodVolatilityBar;
inputBar[0] = 1.0d / forward / periodVolatility * dPlusBar
    + numeraire * omega * nPlus * priceBar;
return new DoubleDerivatives(price, inputBar);
}
```

If you run that code and put a break-point before returning the result, you will notice that for any input, the value of the variable dPlusBar is always 0. That is where the subject matter expert eye can make a difference with respect to an automatic differentiation – see Chapter 4 for more on automatic differentiation, including for the Black formula. What is the d_+ variable? This is the number obtained by solving the equation indicating for which value of the underlying normal random variable, the strike is achieved. Take the example of a call. If the option is exercised at a lower value, we exercise in some case with a loss; if the option is exercised only for a higher value, we do not take full advantage of the option and do not replicate its full value and we have a 0 profit while we could have a positive one. In conclusion, the value d_+ is the optimal value at which the exercise boundary should be placed. Optimal for a one dimensional function in the interior of its domain means – subject to differentiability – that the derivative is 0. This is exactly the impact of this theoretical result that we see in practice. The derivative of the price – the output – with respect to the variable d_+ is always 0; locally the final result is not sensitive to changes in d_+. That derivative, in our code implementation, is the variable dPlusBar.

With our subject matter expert knowledge applied to the above code, we can remove the variable dPlusBar from the code and remove all the code referring to it, replacing it by 0. This required human eyes and subject expertise; it is very difficult to reproduce in an automated way. Implementing those changes, we obtain the code of Listing 3.3.

Listing 3.3 Implementation of the Black 76 formula in adjoint mode with code optimization

```
DoubleDerivatives price_Aad_Optimized(double forward,
    double volatility, double numeraire, double strike,
    double expiry, boolean isCall) {
// Forward sweep - function - see above
...
// Backward sweep - derivatives
double priceBar = 1.0;
double nMinusBar = numeraire * omega * -strike * priceBar;
double dMinusBar = NORMAL.getPDF(omega * dMinus)
    * omega * nMinusBar;
double periodVolatilityBar = -1.0d * dMinusBar;
double[] inputBar = new double[5];
inputBar[4] = volatility * 0.5 / sqrtExpiry
    * periodVolatilityBar;
inputBar[3] = numeraire * omega * -nMinus * priceBar;
inputBar[2] = omega * (forward * nPlus - strike * nMinus)
    * priceBar;
inputBar[1] = sqrtExpiry * periodVolatilityBar;
inputBar[0] = numeraire * omega * nPlus * priceBar;
return new DoubleDerivatives(price, inputBar);
}
```

With a little bit of thinking, we have been able to reduce the number of lines of the code. Does that reduction translate into a significant impact on the performance? The data to answer to this question is displayed in Table 3.1.

We now move to the performance analysis of the different implementations. The Black function and the three AD implementations described above are run one million times each. The AD implementations provide the value itself and the five derivatives as outputs. The first implementation is a standard AD version. The code is not detailed here but can be found in the open source code repository associated to the book. The two next implementations are the adjoint AD versions described above. Note that in the code, there are two implementations of the Black function and its AD versions: one called BlackFunction where the *Apache Commons Mathematics Library*[1] is used as underlying mathematical library and a second one called BlackFormula2 where the *Colt mathematics library*[2] is used. The mathematical library is used for the cumulative normal distribution function and the

---

[1] http://commons.apache.org/proper/commons-math/
[2] https://dst.lbl.gov/ACSSoftware/colt/index.html

Table 3.1 Computation time for different implementations of the Black function and five derivatives (with respect to forward, volatility, numeraire, strike, and expiry)

Repetitions	Computation	Time	Ratio
100,000	Black function	75	1.00
100,000	Black function + SAD	230	3.07
100,000	Black function + AAD	180	2.40
100,000	Black function + AAD Optimized	125	1.67

Figures in milliseconds. Each repetition is with 5 data sets and call/put so that there is actually 1,000,000 repetitions.

density function. The figures reported in Table 3.1 are the one obtained with the Colt version.

The first point to notice is that for all versions, the performance ratio is below or close to three. A finite difference implementation, even the fastest one, would have a ratio of at least 6. Any AD implementation improves on any finite difference version. Now you may point out that what is often required is not all the sensitivities, but only a limited number of them. In the case of the Black formula, you may be interested only in the price and the hedging ratio for delta hedging. Note that computing those numbers requires two pricing, this is a computation time ratio of 2. The optimized adjoint version has a ration of 1.67! Even for the case you need only one sensitivity, computing all of them by AD is faster than only one by finite difference.

The optimized adjoint version improves significantly on the non-optimized adjoint version, with the computation time ratio decreasing from 2.40 to 1.67. This may appear as a magical one-off trick. It is actually a very generic technique that we will make more formal when we discuss calibration in later chapters. Moreover, a lot of explicit pricing mechanism in quantitative finance are based on Black-like formulas. The same improvement will be welcome when we discuss SABR pricing by implied volatility in the next section. It also apply to swaption pricing in Hull-White one factor model (see Henrard 2003), valuation of quality option for bond futures, and other cases where an exercise boundary plays a role.

3.2 SABR

The SABR acronym stands for Stochastic Alpha Beta Rho model (see Hagan et al. (2002)). This is a stochastic volatility model often used in interest rate modeling. The parameters of the model are the starting volatility level α, the elasticity coefficient β, the correlation ρ between the forward Brownian motion W_t^1 and the volatility Brownian motions W_t^2 and the volatility of volatility ν.

The equations for the model are

$$dS_t = \alpha_t (S_t)^\beta dW_t^1$$
$$d\alpha_t = \nu \alpha_t dW_t^2$$

with covariance $[W^1, W^2]_t = \rho t$.

A special feature of one of the implementation of the model makes it very suitable as a starting point of AD analysis. In the Hagan et al. (2002) implementation, the price of a call or put in this model is approximated by a Black formula, as described in the previous section, with an implied volatility provided by the formula below. The option strike is denoted K, the expiry θ, and the current forward F. The implied volatility is given by

$$
\sigma(F,\alpha,\beta,\rho,\nu,K,\theta) = \frac{\alpha}{(FK)^{(1-\beta)/2}\left(1 + \frac{(1-\beta)^2}{24}\ln^2\frac{F}{K} + \frac{(1-\beta)^4}{1920}\ln^4\frac{F}{K}\right)}\frac{z}{x(z)}
$$
$$
\left(1 + \left(\frac{(1-\beta)^2}{24}\frac{\alpha^2}{(FK)^{1-\beta}} + \frac{1}{4}\frac{\rho\beta\nu\alpha}{(FK)^{(1-\beta)/2}} + \frac{2-3\rho^2}{24}\nu^2\right)\theta\right)
$$
$$
\tag{3.2}
$$

where

$$
z = \frac{\nu}{\alpha}(FK)^{(1-\beta)/2}\ln\frac{F}{K}
$$

and

$$
x(z) = \ln\left(\frac{\sqrt{1-2\rho z + z^2} + z - \rho}{1-\rho}\right).
$$

The pricing formula is written by using an implied volatility approach. The number computed above has to be used as implied volatility for the Black formula from the previous section. It is an example of "wrong number in the wrong formula" to obtain the correct price. Independently of what is wrong and what is correct in that approach, it is a perfect example of composition on which the AD framework is based and for which it excels. The pricing formula in SABR is

$$
\mathrm{SABR}(F,\alpha,\beta,\rho,\nu,P,K,\theta) = \mathrm{Black}(F,\sigma(F,\alpha,\beta,\rho,\nu,K,\theta),P,K,\theta).
$$

The implementation of the formula (3.2) is straightforward and reproduced in Listing 3.4. The main point of attention I want to emphasize here is the ratio $z/x(z)$ when z is close to 0, this is when the forward F is close to the strike K. The ratio is of the type "0/0" and converge to 1 at that point. The singularity at that point has no financial significance, it is just an artifact on the way the approximation is obtained and the function should be replaced by its limit for that singular point.

As mentioned in the Principles chapter, one has to be careful about branching as the theoretical differentiability is not guaranteed on the branch boundaries. Here the branching is very special as it is a branching for only one point, when $z = 0$ – in practice when z is close to 0. Actually there is more trouble as the limit in $z = 0$ is of the type "0/0." The phenomenons on numerical instability we discussed in

Listing 3.4 Implementation of the SABR implied volatility approximation

```
private static final double Z_RANGE = 1.0E-6;
double volatility(double forward, double alpha, double beta,
      double rho, double nu, double strike, double expiry) {
  double beta1 = 1.0d - beta;
  double fKbeta = Math.pow(forward * strike, 0.5 * beta1);
  double logfK = Math.log(forward / strike);
  double z = nu / alpha * fKbeta * logfK;
  double zxz;
  double xz = 0.0d;
  double sqz = 0.0d;
  if (Math.abs(z) < Z_RANGE) { // z~0, 1st order approx. for x/x(z)
    zxz = 1.0d - 0.5 * z * rho;
  } else {
    sqz = Math.sqrt(1.0d - 2.0d * rho * z + z * z);
    xz = Math.log((sqz + z - rho) / ( 1.0d - rho));
    zxz = z / xz;
  }
  double beta24 = beta1 * beta1 / 24.0d;
  double beta1920 = beta1 * beta1 * beta1 * beta1 / 1920.0d;
  double logfK2 = logfK *logfK;
  double factor11 = beta24 * logfK2;
  double factor12 = beta1920 * logfK2 * logfK2;
  double num1 = (1 + factor11 + factor12);
  double factor1 = alpha / (fKbeta * num1);
  double factor31 = beta24 * alpha * alpha / (fKbeta * fKbeta);
  double factor32 = 0.25d * rho * beta * nu * alpha / fKbeta;
  double factor33 = (2.0d - 3.0d * rho * rho) / 24.0d * nu * nu;
  double factor3 = 1 + (factor31 + factor32 + factor33) * expiry;
  return factor1 * zxz * factor3;
}
```

Section 1.5 applies here also. It is not enough to simply branch for the value 0, we need a protection around it to avoid a picture like the one provided in Fig. 1.2. In our code, we have selected – arbitrarily – the range to be Z_RANGE = 1.0E-6. In that range the value will not be computed by the $z/x(z)$ formula but by an explicit approximation.

Unfortunately this introduces another problem. If we simply fill the range with the constant 1.0, which is the limit in 0, we have a branch for which the function $z/x(z)$ is a constant function of z and thus its derivative is 0. We get an incorrect derivative at one of the most important point, the at-the-money (ATM) strike. A way around this is to not only incorporate the value 1.0 at that point but also its first order approximation. In the branch we use the function $1 - \rho/2z$ as an approximation of the value of $z/x(z)$. In that way we have the correct first order derivative with respect to z at that point. There will still be a slight jump at the branching in $z=1.0E-6$, but at least the shape is correct to the first order.

With this branching in mind, the code for the adjoint AD version is slightly more complex that initially hoped for but still very manageable. The branching will appear also in the backward sweep, but for the rest there is nothing special in this code. The highlights of the code are provided in Listing 3.5. As usual the full code is provided in the open source repository associated to the book.

Listing 3.5 Implementation of the SABR implied volatility approximation: adjoint AD version

```
DoubleDerivatives volatility_Aad(double forward, double alpha,
    double beta, double rho, double nu, double strike,
    double expiry) {
// Forward sweep - function - see above
// Backward sweep - derivatives
double volatilityBar = 1.0d;
double factor3Bar = factor1 * zxz * volatilityBar;
...
double zBar;
double xzBar = 0.0d;
double sqzBar = 0.0d;
if (Math.abs(z) < Z_RANGE) {
  zBar = 0.5 * rho * zxzBar;
} else {
  xzBar = -z / (xz * xz) * zxzBar;
  sqzBar = xzBar / (sqz + z - rho);
  zBar = zxzBar / xz;
  zBar += xzBar / (sqz + z - rho);
  zBar += (-rho + z) / sqz * sqzBar;
}
double logfKBar = nu / alpha * fKbeta * zBar;
...
inputBar[5] += 0.5 * beta1 * fKbeta / strike * fKbetaBar;
inputBar[5] += -logfKBar / strike;
inputBar[6] += (factor31 + factor32 + factor33) * factor3Bar;
return new DoubleDerivatives(volatility, inputBar);
}
```

All the discussion above concerns the implied volatility part. We now have to price the option through composition. This is done in Listing 3.6. We have directly provided the adjoint AD version. The value part is made of two lines, the computation of the implied volatility in the first and the computation of the price based on the Black formula using the volatility as an input in the second.

The adjoint AD is almost as simple, the volatility derivative is provided by the adjoint AD version of the Black formula price. Then the sensitivities to the different inputs are collected from the derivatives of the two functions used in the pricing. When the composition is used not only for each line of code but for the call to intermediary methods, the derivatives are not derivatives of elementary functions anymore, but the derivatives provided by the AD versions of the lower level methods.

Listing 3.6 Implementation of the SABR price through implied volatility

```
DoubleDerivatives price_Aad(double forward, double alpha,
    double beta, double rho, double nu, double numeraire,
    double strike, double expiry, boolean isCall) {
  DoubleDerivatives volatility = SabrVolatilityFormula
    .volatility_Aad(forward, alpha, beta, rho, nu,
    strike, expiry);
  DoubleDerivatives price = BlackFormula.price_Aad_Optimized(
    forward, volatility.value(), numeraire, strike,
    expiry, isCall);
  double priceBar = 1.0d;
  double volatilityBar = price.derivatives()[1];
  double[] inputBar = new double[8];
  inputBar[7] += price.derivatives()[4] * priceBar;
  inputBar[7] += volatility.derivatives()[6] * volatilityBar;
  inputBar[6] += price.derivatives()[3] * priceBar;
  inputBar[6] += volatility.derivatives()[5] * volatilityBar;
  inputBar[5] += price.derivatives()[2] * priceBar;
  inputBar[4] += volatility.derivatives()[4] * volatilityBar;
  inputBar[3] += volatility.derivatives()[3] * volatilityBar;
  inputBar[2] += volatility.derivatives()[2] * volatilityBar;
  inputBar[1] += volatility.derivatives()[1] * volatilityBar;
  inputBar[0] += price.derivatives()[0] * priceBar;
  inputBar[0] += volatility.derivatives()[0] * volatilityBar;
  return new DoubleDerivatives(price.value(), inputBar);
}
```

Note that at this stage, when we are running the code of the external function – SABR price in our case we don't need to know what happened in the internal computation of the underlying code. How the Black formula price is computed is irrelevant, only the final results of that method – the value and its derivatives – and the memory space to hold them is relevant. This will not be the case for the automatic AD by operator overloading described in Chapter 4. The standard implementation in that case uses a *tape* that records every elementary operation, including those inside the lower level methods. The recorded information is reused at a later stage when the derivatives are computed. In some sense the tape creates an in memory explicit version of the computation as a SAC list. The memory impact of automatic adjoint AD is one of its main drawbacks.

Like for the Black formula, we can check the performance of the different implementations of the SABR price with approximated implied volatility. The number of inputs is now eight. The results are presented in Table 3.2. The results were obtained by running the analysis code `SabrFormulaAnalysis`.

Again the performance is quite impressive with the computation of the value and the 8 derivatives taking less than twice the computation time of one value. As for Black formula, computing one sensitivity by finite difference would be slower than computing all the 8 sensitivities by AD.

Table 3.2 Computation time for the SABR price and eight derivatives (with respect to forward, alpha, beta, rho, nu, numeraire, strike, and expiry)

Repetitions	Computation	Time	Ratio
100,000	SABR price	240	1.00
100,000	SABR price + AAD Optimized	440	1.83

Figures in milliseconds. Each repetition is with 5 data sets and call/put so that there is actually 1,000,000 repetitions.

Note that in practice this formula is often used with interest rate products. It means that the inputs of the above formula, like the forward rate F and the numeraire P are themselves output of previous methods taking 10–100 inputs. In practice the saving are compounded with the saving described in the next section. This accumulation of saving multipliers make the technique extremely efficient, even for vanilla interest rate products like swaptions.

3.3   Coupon Sensitivities

Where we encounter sensitivities that are financially useful but ADly complex to extract, sensitivities that have weak financial meaning and ADly complex to extract and finally sensitivities that are financially useful and directly available in all AD implementations.

In this section, we analyze interest methods related to present value and its derivatives for interest rate swaps. In the finance jargon, the derivatives are often called *sensitivities*, we use that name in this section. The sensitivities are to the curves – discounting and forward – used to value the swaps. The meaning of sensitivities to curves is explained below. The instruments we analyze are composed of fixed coupons and Ibor coupons. We work in a multi-curve framework as described in Henrard (2014). Here we only sketch the main financial problem and the important definitions. Readers unfamiliar with the multi-curve framework fascinating world are referred to the above reference for more details.

The challenges would not be very different in a simpler one curve framework, which was the main interest rate framework for derivatives before the crisis, but we prefer to present a more recent framework. In term of efficiency, the AD implementation appears even more spectacular in the multi-curve framework as there are more inputs and more sensitivities to compute. In a typical multi-curve framework for one currency, there would be four or five curves, each calibrated to 10–25 instruments. A typical simple vanilla single currency interest rate book will exhibit sensitivities to 50–125 inputs.

The first challenge when working with curves is the dimensionality. In the theoretical description of our function by Eq. 2.3 and the code description in Table 2.1, all the functions have a finite number of arguments. From a theoretical point of view a curve $f : [0, T] \rightarrow \mathbb{R}; t \mapsto f(t)$, like the one used to represent the discount factors, is an element of an infinite dimensional space, the space of functions from an interval to $\mathbb{R}$. Obviously the element of the space is represented in the code by a finite number of data that are used to construct the element of the infinite dimensional space. The usual representation of a curve is through the nodes of an interpolated curve or the parameters of a functional curve. But formally the problem we are facing involves a function from a finite number of data to an element of an infinite dimension space composed with a function from the infinite dimension space to the one dimensional present value. We will need some mechanism to deal with this infinite dimensional space.

Before developing the algorithm differentiation challenges further, we describe the notations and main features of the multi-curve framework. Let $P_X^D(v)$ be the discount factor to the valuation time in currency X for a payment at time v. The present value of a fixed coupon paying in time v a rate r on a notional N with accrual factor δ is given by

$$\text{PV} = N P_X^D(v) \delta r.$$

The discount factor function is a function $P_X^D : [0, T] \rightarrow \mathbb{R}; t \mapsto P_X^D(t)$.

We call a j-Ibor floating coupon a financial instrument which pays at the end of the underlying period the Ibor rate set at the start of the period. The details of the instrument are as follow. The rate is set or fixed at a date t_0 for the period $[u, v](t_0 \le u < v)$ of length equal to the tenor of the Ibor index j, at the end date v the amount paid in currency X is the Ibor fixing $I_X^j(t_0)$ multiplied by the conventional accrual factor δ for the period. The lag between t_0 and u is called the *spot lag* or *settlement lag*. The difference between u and v is the tenor of the index j. All periods and accrual factors should be calculated according to the day count, business day convention, calendar and end-of-month rule appropriate to the relevant Ibor index.

The present value of the Ibor coupon paying in v the rate fixed in t_0 for the period $[u, v]$ on the notional N for an accrual factor δ is given by

$$\text{PV} = N P_X^D(v) \delta F^j(t_0, u, v)$$

where $F^j(t_0, u, v)$ is called the forward rate for index j viewed from the valuation date. The forward rate function is a function $F^j : [0, T] \rightarrow \mathbb{R}; t \mapsto F^j(t, u(t), v(t))$. The dates u and v are duplicate information, as they can be inferred from t and the convention of the index j. They are usually provided in literature related to interest rate for the reader convenience but could be simply ignored. Looking at the multi-curve foundations, the function depends only of the fixing date and is thus a one dimensional function.

We have described the two main ingredients of an interest rate swaps: the fixed coupons and the Ibor coupons. A full swap is simply a combination of those two types of coupons, one type on each leg. Its total present value is obtain as the sum of the present values of the different coupons. The "sum" part does not cause any problem for the AD framework.

Before starting the implementation, and in particular to review how to work around the infinite dimension problem in practice, we have to ask ourself "What are the derivatives (sensitivities) that a risk manger would like to see in such circumstances?" I can see three different, and non-exclusive, answers to that question.

The first possible answer is that we would like to see the sensitivity to each discount factor and forward rate. This would be the sensitivity for each discount factor at time v at which there is a payment and for each forward rate at time t_0 at which there is a fixing. This is where the infinite dimensional question enter into play. You cannot create a vector of all possible times and fill the sensitivity to the relevant ones with a number and leave the rest at 0. There is no exhaustive list of times for which one may want to compute this, a priori any time between 0 and T is a possible payment or fixing time.

One can claim that in practice, the fixing and payment dates are only precise to the date level, so the values are not any real number in the interval $[0, T]$ but in a discrete subset of it representing each day in the period. Even if we take only good business days into account, for a 50-year curve, this means a little bit more than 12,500 potential times. Would you like to carry around an array of that size when you have only a couple of fixings or payments? This would probably be a huge cost in memory management. Moreover the argument may be valid for payment time, but when we will move to options, there is a clear expiry time and one would like to model this down to the hours and minutes. An option with 23h00 to go is not the same as an option one hour after expiry! Often the implied volatilities used with Black-like formulas are time dependent and there a full continuous time dimension is required. The mechanism we put in place to deal with the infinite dimension of the interest rate curve – or finite with very large dimension – will be useful in other circumstances later.

In conclusion, due to the very large discrete potential times or due to the really continuous feature of the time, we may want to represent the sensitivities not through an array of fixed length. We have to move away from the simplified framework of the AD we have worked with until now and accept that some information is not simply represented by a double or an array of them. Our suggestion in the case of interest rate sensitivities is to use lists, i.e. a collection without an a priori size. But list of what? In one of our implementations[3], we have used the objects `ZeroRateSensitivity` for sensitivities associated to discount factors and `IborRateSensitivity` for sensitivities associated to forward Ibor rates.

By tradition, instead of representing the sensitivity in term of sensitivity to the discount factor, it is represented as the sensitivity to the zero-coupon rate $r(t)$

---

[3] The implementation is the *Strata* framework developed by OpenGamma. The code is available under an OpenSource license at https://github.com/OpenGamma/Strata.

associated to it by $P_X^D(t) = \exp(-r(t)t)$. The sensitivity object for one cash flow is represented in Listing 3.7[4].

Listing 3.7   Object for sensitivities to discount factors

```
public class ZeroRateSensitivity implements PointSensitivity {
  Currency ccyDiscount;
  LocalDate date;
  double value;
  Currency ccySensitivity;
  ...
}
```

Note that there are two currencies, one for the currency in which the discounting is performed, i.e. the X in the formula and one for the unit in which the sensitivity is represented. There is good argument to claim that one currency would be enough for discount factor sensitivity and I'm ready to accept those arguments, even if in some cases you may want to express all the results in a common currency or convert from one currency to another.

But in the case of the sensitivity to forward Ibor rate, which is the object represented in Listing 3.8, one definitively need two information: the index of the forward and the currency in which the amount is paid. Even if a currency is implicit in the index, we need a second currency to know in which unit the payment is done. This is important for quanto-like payments in particular.

Listing 3.8   Object for sensitivities to forward rates

```
public class IborRateSensitivity implements PointSensitivity {
  IborIndex index;
  LocalDate fixingDate;
  Currency currency;
  double sensitivity;
  ...
}
```

With each of those objects we can store the sensitivity to one discount factor or one forward rate. With a list of them, we can store as many sensitivities as necessary. An object collecting the sensitivities is summarized in Listing 3.9.

Note that at this stage the data is stored in a `List`, not an array. This is important to note in relation to our remarks above. There is no fixed size array for all the potential dates of a point sensitivity. The length of the `PointSensitivities`

---

[4] The implementation used in practice is slightly more complex than the one described in this text. In particular the actual implementation has features for serialization, builders and combination of objects. We emphasize here only the AD related features.

Listing 3.9 Object for point sensitivities in the multi-curve framework

```
public class PointSensitivities {
  List<PointSensitivity> sensitivities;
  . . .
}
```

embedded list depends on the instrument priced and potentially the model and calibration procedure used.

With the information in that object, we have enough data to answer the questions like, to which date do I have discounting factor sensitivity or to which date do I have fixing risk. In terms of market risk, to my opinion, the most important is the fixing date sensitivity. As a fixing is a daily one-off there is limited diversification one can do, it is not possible to take the fixing from another source, like one can borrow from another counterpart, one cannot really postpone the fixing to the next day, while is it often possible to borrow overnight. On the other side, knowing the exact profit impact of each daily fixing would allow to foresee the impact of fixing manipulation. But I, naively, cannot believe that one would ever use AD for a such evil purpose as market manipulation.

We call the first level of sensitivity, as described above, the *point sensitivity*. Each potential point on the curves can produce a sensitivity. But this type of sensitivity would not be produced by an Algorithmic Differentiation sensu stricto. The sensitivities or derivatives computed are not with respect to an input but to an intermediary value, they can nevertheless be very useful and should not be excluded from a general financial library. Often in libraries where the sensitivities are produced by finite difference or automatic AD, those point sensitivities are unfortunately not available. In the case of finite difference they are not even computed, and in the case of automatic AD they may have been computed and recorded in the tape, but finding them may be more complex as they are not earmarked. We will come back to similar issues in Chapter 4 on automatic AD and in Chapter 5 on adaptation of AD to compute numbers closely related to derivatives.

The second level of sensitivities we produce is called *parameter sensitivity* or *internal representation sensitivity*. This is the sensitivity to the internal parameters used to represent the calibrated curves. The most commonly used curve descriptions is through interpolated zero-coupon curves, but many other representation are possible. Some useful representations are described in (Henrard 2014, Chapter 5). This sensitivity type is probably, or should be, the least useful of the three we describe here. It depends on the internal representation of the curve in the library. This internal representation is not the most useful to the business users. A curve can be represented by zero-coupon rates or discount factor; by changing the interpolation mechanism, both representations can produce exactly the same result. The internal representation would be invisible to an outside user. Providing the parameters sensitivity to the end user is providing number potentially widely different for the same inputs and same outputs; this can be surprising for a user that does not know about the internal working of the library.

In particular, the sensitivity to zero-coupons rates are often used as a risk measure. Zero-coupon rate are not defined for all curves. In particular forward Ibor curves described directly through forward rate and not through pseudo-discount factors do not have zero-rate representation. The theoretical description of those curves in the multi-curve framework is provided in (Henrard 2014, Chapter 3). It means that the parameters that are used to describe curves are in general not homogeneous and often don't have a direct intuitive meaning. Relying on those numbers for risk management may be unwise. The zero-coupon PV01 enter, to my opinion, in this category of ill-defined risk measures. Or more exactly is it defined only in a restricted set of curve description. Relying on them would restrict the risk manager view of the risk world.

The algorithmic differentiation meaning of the parameter sensitivities is clear, it is the derivatives of the final present value with respect to the parameters used to internally store the curve. It is the derivative to an intermediary variable. Like for the point sensitivities, how to extract the information computed as some stage of the process will depend on the implementation. But what the external business user can do with those numbers dependent of the internal representation and not of financial information is unclear.

In implementation terms, the parameter sensitivity can be computed as a *projection* of the point sensitivity to the parameters of the curves. By projection, we mean here multiplying the partial derivative of one of the functions composing the algorithm by the derivative of the remaining composed functions with respect to the intermediary variables used. A version of that approach can be found in the class `AbstractRatesProvider` of the OpenGamma's Strata library. This is a generic algorithm that applies to any point sensitivity. This part of the code is not instrument specific but curve description specific. Only a unique implementation in a library is required.

One of the important practical aspect of the parameter sensitivity is the object containing the sensitivity. From a pure data point of view, the result is simply a long vector; from a business point of view it is important to know to which curve and point each number refers and in which currency it is expressed. Moreover, the number of curves and currencies to which an instrument present value depends is a priori unknown. If the data is stored as a unique vector, it would need to be a vector with all the curves used in the system, not only the one related to the instrument analyzed. This would mean again a huge quantity of 0 stored. For that reason we suggest to store the parameter sensitivity in an object composed of a list of sensitivities to a single curve. Each single curve sensitivity contains the actual sensitivity vector, the currency in which the sensitivity is expressed and the curve metadata, i.e. its name and the description of the parameters. An example of such an object is provided in Listing 3.10.

The third level of sensitivity is the *market quote sensitivity*. Another term often used for those sensitivities is *par rate sensitivity*. I prefer to use the market quote terminology, as some instruments are quoted through par rate but others are quoted using other mechanism as spread, futures price or bond price. The par rate sensitivity term appears more restrictive and we would like to have an approach which is as generic as possible.

Listing 3.10   Object for parameter sensitivities in the multi-curve framework

```
public class CurveCurrencyParameterSensitivities {
  List<CurveCurrencyParameterSensitivity> sensitivities;
  . . .
}

public class CurveCurrencyParameterSensitivity {
  CurveMetadata metadata;
  Currency currency;
  double[] sensitivity;
  . . .
}
```

Computing the market quotes sensitivities means that the curve have been calibrated from market quotes. The curve calibration process is discussed in more details in Section 6.1. One of the output of that procedure is obviously the calibrated curves, i.e. the set of internal parameters describing the curves. Another very important output is the *transition* or *Jacobian* matrices. Those matrices are the partial derivatives of the internal parameters of the curves with respect to the market quotes of the instruments used to calibrate the curves.

With the description above, the numerical procedure to obtain the market quote sensitivities is direct. The derivative of the present value (or any other measure) with respect to the market quotes is the multiplication of the transition matrix by the parameter sensitivity computed at the previous step:

$$\frac{\partial PV}{\partial q_i} = \sum_j \frac{\partial PV}{\partial p_j} \frac{\partial p_j}{\partial q_i}.$$

Like for the parameter sensitivity, one of the challenges to the market quotes sensitivities is the object representing it. We suggest to use the same object than the one described in Listing 3.10. Even if the business description of the two results are not the same, the structure of the data is the same. The data has the same format, but the metadata will distinguish between the two types of sensitivities. For several curves, there is a sensitivity, represented by an array, to data in a given currency.

The object containing the *transition matrices* is not trivial. Each calibrated curve potentially depends on several market quote curves. The different curves can have different sizes. The order in which the curves are calibrated is itself important. An example of an object containing a representation of the transition matrices for one curve is proposed in Section 6.1 on curve calibration.

The performance obtained by this approach is presented in Table 3.3. In all cases, the sensitivities are computed in a time which is less than five times the time of the present value. The time presented in the table does not include the curve calibration.

Table 3.3 Computation time for the present value and different type of sensitivities. The multi-curve set has two curves and a total of 30 nodes

Repetitions	Computation	Time	Ratio
1,000	Swaps present value	195	1.00
1,000	Swaps point sensitivity	280	1.44
1,000	Swaps (point + parameter) sensitivity	715	3.67
1,000	Swaps (point + parameter + quotes) sensitivity	900	4.62

Figures in milliseconds. Each repetition is with three swaps with maturities 5, 10 and 30 years.

3.4 Monte Carlo

Monte Carlo pricing of financial derivatives is certainly a numerically challenging problem in quantitative finance. But the Algorithmic Differentiation version of it does not present any specific challenges on top of the pure pricing challenges. Obviously the different steps have to be treated carefully but they can be seen as a quite direct application of the AD methodologies.

A Monte-Carlo pricing for a European option can be described in several steps:

Random number Generate a set of random numbers $r_{f,s}$ where f $(1 \leq f \leq F)$ represents a factor of the model and s $(1 \leq s \leq S)$ a scenario.

Scenario From each scenario, generate the market at the expiry date θ. The scenario market will depend on the model parameters V_p $(1 \leq p \leq P)$ and the underlying $M_{0,u}$ $(1 \leq u \leq U)$ - yield curves for example - in a given numeraire

$$M_{\theta,s} = g_1(r_{.,s}, V, M_0).$$

where $M_{\theta,s} = (M_{\theta,s,u})_{u=1,...,U}$ is a market vector for each scenario and $g_1 : \mathbb{R}^F \times \mathbb{R}^P \times \mathbb{R}^U \to \mathbb{R}^M$. Through the stochastic model, the market description is obtained from the initial market and the model parameters. The market is for example the full yield curves in an interest rate term structure model.

Numeraire and pay-off From the different markets, compute the pay-off and the numeraire

$$P_{\theta,s} = g_2(M_{\theta,s}) \quad \text{and} \quad N_{\theta,s} = g_3(M_{\theta,s}).$$

Average The present value is the average or expectation of the pay-off discounted by the numeraire:

$$PV_0 = N_0 \frac{1}{S} \sum_{s=1,...,S} \left(N_{\theta,s}^{-1} P_{\theta,s} \right).$$

where N_0 is the value in 0 of the numeraire.

The Adjoint Algorithmic differentiation start from the last step. The derivatives are computed with respect to the underlying parameters M_0 and the model parameters V

Average There is one sensitivity for each scenario s:

$$\bar{N}_{\theta,s} = -N_0 N_{\theta,s}^{-2} P_{\theta,s} \quad \text{and} \quad \bar{P}_{\theta,s} = N_0 N_{\theta,s}^{-1}.$$

Numeraire and pay-off For each of the dimension m representing the market and each scenario s we have one sensitivity:

$$\bar{M}_{\theta,s,m} = D_m g_2(M_{\theta,s})\bar{P}_{\theta,s} + D_m g_3(M_{\theta,s})\bar{N}_{\theta,s}.$$

Random number The starting parameters and curves are used in each scenario s and for each dimension of the underlying market m:

$$\bar{V}_p = \sum_{s=1,\ldots,S} \sum_{m=1,\ldots,M} D_{V_p} g_{1,m}(r_{.,s}, V, C)\bar{M}_{s,m}$$

$$\bar{C}_u = \sum_{s=1,\ldots,S} \sum_{m=1,\ldots,M} D_{C_u} g_{1,m}(r_{.,s}, V, C)\bar{M}_{s,m}$$

Even if the final formulas contains several layers of sums, they are very easy to compute provided than the derivatives of g_1, g_2 and g_3 can be computed. Obviously those functions will be instrument and model dependent. But in general they are relatively simple. For standard options, g_2 is often a piecewise linear function and g_3 is often a linear combination of market factors. The function g_1 is usually the most complex part, it describe how to generate future market data from current data and model parameters. But even in a Libor Market Model with the forward market generated by Euler or Predictor-Corrector approach, it can be decomposed in exponentials, divisions, multiplication and additions. The AD development for those functions is usually straightforward, even if tedious.

Bibliography

Bachelier, L. (1900). *Théorie de la Spéculation*. PhD thesis, Ecole Normale Supérieure.

Black, F. (1976). The pricing of commodity contracts. *Journal of Financial Economics*, 3(1–2):167—179.

Hagan, P., Kumar, D., Lesniewski, A., and Woodward, D. (2002). Managing smile risk. *Wilmott Magazine*, Sep, 2002(5):84—108.

Henrard, M. (2003). Explicit bond option and swaption formula in Heath-Jarrow-Morton one-factor model. *International Journal of Theoretical and Applied Finance*, 6(1):57–72.

Henrard, M. (2014). *Interest Rate Modelling in the Multi-curve Framework: Foundations, Evolution and Implementation*. Applied Quantitative Finance. London:Palgrave Macmillan. ISBN: 978-1-137-37465-3.

Chapter 4
Automatic Algorithmic Differentiation

Big brother is watching you: AD tapes – Automatic AD saves programing time, not execution time – Human are still useful to computers – On n'oublie rien de rien. On s'habitue c'est tout

On top of theoretical Algorithmic Differentiation as described in most of this book, there is also the computer science art of *Automatic Algorithmic Differentiation*. This is the art of writing code that performs the Algorithmic Differentiation automatically. The developments required are relatively heavy at the start, but from there on, there should be only a minimal development cost.

Automatic Algorithmic Differentiation is not the main subject this book. In most of the chapters, we discuss AD's principles and how it can be applied to finance. Nevertheless to understand the subject globally a minimum discussion on the automation is required.

Generally, automatic differentiation can be achieved in two ways. One approach is to generate the AD code automatically from the original function code. A second approach is by "operator overloading." The code is almost unchanged but the objects on which it operates are not primitive `double` anymore but *augmented doubles*. The augmented doubles contains the double information on its value and also some extra information required to perform the AD calculations automatically. Those augmented doubles can be created for Standard AD and Adjoint AD. In our implementation, for the standard version, the extra information is the array of already computed derivatives. For the adjoint version the extra information is the index of the operation in an AD *tape*. The notion of AD tape is described later in this chapter. In this book we chose to present the operator overloading approach. To be precise, as we use Java as our example language, we don't actually use operator overloading but operator replacement by methods. This is maybe not as neat when writing/reading the code. But on the other side it is very explicit on which parts are the simple direct code on primitive doubles and which parts are the overloaded versions on augmented doubles. From a didactical point of view it is important to be able to distinguish clearly between them. Operator overloading may appear in Java 10, in which case I would rewrite the code to use the new feature.

Many references regarding the computer science art of automatic Algorithmic Differentiation can be found in the recent book Naumann (2012). The website `http://www.autodiff.org/` can also be used as a starting point for many tools and references.

© The Author(s) 2017
M. Henrard, *Algorithmic Differentiation in Finance Explained*,
Financial Engineering Explained, DOI 10.1007/978-3-319-53979-9_4

In the advanced chapters of this book we argue that Algorithmic Differentiation done by subject matter experts can improve the already amazing performance of Algorithmic Differentiation. The automatic differentiation is certainly a very useful approach and once fully implemented can save a lot of development time. But there is more to algorithmic differentitation than simple and blind application of rules without an expert eye.

Regarding the efficiency of AD with an expert eye on the implementation, we will show in the "at best" calibration cases described in a later chapter, that the derivatives of complex process, like pricing complex derivative with ad hoc model calibration, can be obtained for almost free. The ratio between the price alone and the (numerous) derivative will be barely above 1; in the region of 1.1. To compute the price plus 100 of its derivative, the computation time is the one of the price plus 10%. You get 100 extra numbers in 10% of the time required for the first number! This was not achieve by a generic AD implementation that reproduces in the derivative computation the algorithm of the pricing but a specific AD computation that take into account the structure of the problem.

In Naumann and du Toit (2014), the author states *"Doing AD by hand on any production size code is simply not feasible and is incompatible with modern software engineering strategies."* I disagree with this statement for the following reason. The *philosophy* of AD is to use the composition (see Section 2.2). Each part of the composition need to have a derivative version. It should be mainly irrelevant if each block of the composition is written automatically or manually. Disregarding form the start any implementation approach, it being manual or automatic (or finite difference), as part of a general AD is incoherent with my view of the general AD philosophy. The exclusion of an approach under the authority argument that is not "modern" is not a strong argument. A general AD implementation should accept any implementation for each part of the composition. An implementation that accepts only one specific type of implementation for each part would be a weak one. To have the best team you need to be allowed to take the best element for each position and make sure they work together smoothly, this is the case for AD as for team sport. We will present later examples of mixed implementations involving both manual and automatic AD in the same computation.

One of the drawbacks of fully automatic AD, is that it requires full access to source code, even for third party libraries. It is not enough to have a derivative version of each method; it has to be in the exact same code approach. The manual implementation does not have this problem, different approaches can be combined for each line of code. By extension an approach that combines automatic differentiation with manual glue will allow diversity and efficiency.

Another of the drawbacks of automatic AD is the memory requirements. For example in Naumann and du Toit (2014), the authors states *"Moreover, AD implementations based on operator overloading techniques (such as dco/c++) amplify this problem since they need to store and recover at runtime an image of the entire computation (commonly referred to as the tape) in order to ensure the correctness of the calculation."* Roughly speaking standard automatic AD with operator overloading record each code operation with entries logging the type of operation, the number involved, the value computed and, in a second phase, the derivatives. The record of all the those operations is called the AD tape. At the very end of the computation, that very long list of values and derivatives can be replayed backward to get the derivatives results. In any case, all the steps required for the final computation, even if there are many of them, are kept in memory up to the very end of the process to achieve the derivative computation. The more manual implementation can release memory at each stage of the computation if the intermediary numbers are not used any more. The intermediary garbage can be collected when not used anymore.

The memory drawback is also accompanied by a lost performance. Not only the results and derivatives are recorded but also all the description on how the results are obtained. Manipulating all this data is by itself time consuming. In some of the examples we will compare the time required to do the computation of a value and no derivative with primitive doubles to the time with augmented doubles. This shows the cost of the operator overloading which is not linked to the computation of derivatives itself but to the gathering of data required to do it automatically.

Obviously different implementation will have different costs. As an indicative level, for our implementation of automatic Adjoint AD, performing the computation with the augmented doubles instead of the simple doubles multiply the computation time by a factor of two to five. This is only for the recording of the operation and not the computation of the derivatives. The total cost of the recording and the computation of the derivatives is between two and ten. The theoretical results indicate an upper bound for the ratio around three to four. The implementations by subject matter experts as discussed in different places in this book provide ratios between 1.1 and 4.

A general Automatic Algorithmic Differentiation tool is proposed in Naumann and du Toit (2014). Some part of it are also described in Naumann (2012). The C++ implementation is called dco/c++ (derivative computation by overloading) and is developed at RWTH Aachen University in collaboration with NAG. In the financial library sector, the vendor Fincad also propose a variation of AD that they call Universal Algorithmic Differentiation™. This last implementation being closed source, it is difficult to assess to which extend it differs or overlap with the description done in this book. Another automatic AD implementation is TapeScript[1]. TapeScript is an open source library for adjoint algorithmic differentiation (AAD) developed and maintained by CompatibL

4.1   Standard Algorithmic Differentiation by Operator Overloading

The automatic algorithmic differentiation approach by operator overloading applies the standard code to *augmented objects*, while the manual AD approach applies

---

[1] https://github.com/compatibl/tapescript

augmented code to standard objects. Those augmented objects need to be created carefully.

A potential implementation for an object storing a value and its derivatives, suitable for automatic standard AD, is proposed in Listing 4.1. The code of the object can be found in the GitHub repository described in the introduction in the package `type`. The object is called `DoubleSad`. This is an augmented `double` to be used in Standard Algorithmic Differentiation. In standard AD, at each step the derivatives with respect to each input is computed. The number of inputs is known and an array of the correct size can be carried around. The computation of the derivatives is done at the same time as the computation of the value itself.

Listing 4.1   Object storing information for Standard Algorithmic Differentiation by operator overloading

```
public class DoubleSad {
  private final double value;
  private final int nbDerivatives;
  private final double[] derivatives;

  public DoubleSad(double value, double[] derivatives) {
    this.value = value;
    this.derivatives = derivatives;
    nbDerivatives = derivatives.length;
  }
  ...
  static public DoubleSad[] init(double[] inputs) {
    int nbInputs = inputs.length;
    DoubleSad[] init = new DoubleSad[nbInputs];
    for(int loopi = 0; loopi < nbInputs; loopi++) {
      double[] initDot = new double[nbInputs];
      initDot[loopi] = 1.0d;
      init[loopi] = new DoubleSad(inputs[loopi], initDot);
    }
    return init;
  }
}
```

The first step in each computation is to initialize the different `DoubleSad` with the input values and the relevant derivatives. The derivatives are trivial as the derivative of an input with respect to the input itself is 1.0 and with respect to the other inputs is 0.0. This initialization is done by the `init` method.

For each elementary operation, like addition, subtraction, multiplication, sinus, exponential, power, normal cumulative density, etc. we have to explain how the operation acts on the value and on its derivatives. This is done through the different static methods with the name of the operation stored in the `MathAad` class. Some of the method are provided in Listing 4.2. For example the addition is described by the `plus` method. The addition of two `DoubleSad` is done by adding the values, which are simple doubles and by adding the derivatives arrays element by element.

Similar descriptions are required for each operation (subtraction, multiplication, sine, cosine, exponential, logarithm, square root, etc.). The code is not presented for the other operation, as it would be a very long listing without real didactic value. The full code can be found in the open source repository.

Listing 4.2   Mathematical operations for Standard Algorithmic Differentiation by operator overloading

```
public class MathSad {

  public static DoubleSad plus(DoubleSad d1, DoubleSad d2) {
    int nbDerivatives = d1.getNbDerivatives();
    double valueOutput = d1.value() + d2.value();
    double[] derivativesOutput = new double[nbDerivatives];
    for (int loopd = 0; loopd < nbDerivatives; loopd++) {
      derivativesOutput[loopd] =
          d1.derivatives()[loopd] + d2.derivatives()[loopd];
    }
    return new DoubleSad(valueOutput, derivativesOutput);
  }
  ...
}
```

Using that object, the code for the initial test function described by Listing 2.1 in the previous chapter can be adapted easily. The code for the automatic Standard Algorithmic Differentiation version is given in Listing 4.3. With respect to the original code, the operator – like + – have been replaced by the method of the similar name and the mathematical functions – like exponential – have also been replaced by a method of similar name but in the MathSad class instead of the Math class. For the rest the code looks very similar.

Listing 4.3   Simple function code by Automatic Standard Algorithmic Differentiation

```
static public DoubleSad f_Sad_Automatic(DoubleSad[] a) {
  DoubleSad b1 = MathSad.plus(a[0], MathSad.exp(a[1]));
  DoubleSad b2 = MathSad.plus(MathSad.sin(a[2]), MathSad.cos(a[3]));
  DoubleSad b3 = MathSad.plus(MathSad.pow(a[1], 1.5d), a[3]);
  return MathSad.plus(MathSad.multipliedBy(MathSad.cos(b1), b2), b3);
}
```

As the algorithm is progressing through the code, the value is computed and stored in the value part of the DoubleSad. The derivatives are computed at the same time and stored in the derivatives array part of the augmented double.

Performance results of this implementation will be provided later in the chapter, in particular in Table 4.2, when we compare them with different versions of manual and automatic Adjoint AD.

4.2   Adjoint Algorithmic Differentiation by Operator Overloading

The implementation of the adjoint mode is usually technically more complex. The reason is that each line of code is analyzed twice, once in the forward sweep where the value is computed and once in the backward sweep where the derivatives are computed. It means that all the operations done in the original code through the forward sweep have to be recorded and to be replayed later. This recording of operations is not only at the level of one method but through the full code, between the different methods. If two methods are composed, the recording will contain the full list of all operations in both methods. The accumulation of basic operations span the full code run.

The list of recorded operations is usually called the *tape*. The tape contains the list of all elementary operations performed and enough information to know to which variables the operations applied. The unit for the recording is not the *line of code* as in the high level language but at the level of more fundamental operations within one line of code. Each operation acts on 0, 1 or 2 variables. In our example implementation, we have implemented the following fundamental operations: IN-PUT, MANUAL, ADDITION, ADDITION1, SUBTRACTION, MULTIPLICATION, MULTIPLICATION1, DIVISION, SIN, COS, EXP, LOG, SQRT, POW, POW1, NORMALCDF. For most of them, the name is explicit enough and they do not require more explanation. The ones requiring more explanations are explained below.

INPUT: The fundamental variables with respect to which the derivatives need to be computed, this is the starting point of the recording and at the same time the end point of the backward sweep. Once the backward algorithm reaches one of those record, nothing else need to be done with that information, the required derivative is obtained. The tape will keep the records available for future extraction.

MANUAL: This is the marker in the automatic AD to indicate non-automatic computations, i.e. that some computation of derivatives have been done in a different way to the current implementation of automatic AD and the result obtained with a different format is fed back into the tape. To my opinion, the main principles of AD is composition of functions and its derivatives. A composition can be achieved by combining adjoint, standard, automatic or manual implementation. Those different approaches should not be exclusive from each other. We will come back to this special operation when analyzing mixed implementations later in the chapter.

ADDITION1, MULTIPLICATION1 and POW1: When adding, multiplying or taking the power of numbers, two possibilities arise. If the two numbers are *augmented doubles*, the derivatives with respect to both is required. On the other side, if one of the number is a constant, not dependent of the inputs, the derivative with respect to only one of them is required. The "1" versions of the three operations above are to be used if only the first element is a augmented double and the other one is a primitive double – a constant for derivatives computation purposes.

The operations for the adjoint AD are performed on augmented doubles that we call `DoubleAad` in the code associated to the book. A shorten version of the code, where only a couple of the ingredients are provided, can be found in Listing 4.4. The full code can be found on the GitHub repository. The information recorded in that object is the value itself and the *index, virtual address* or *recording number* in the tape where the operations that led to that result are located. The number called *tape index* in this book is also called *virtual address* in the literature.

Listing 4.4 Object storing information for Adjoint Algorithmic Differentiation by operator overloading

```
public class DoubleAad {
  private final double value;
  private final int tapeIndex;

  public DoubleAad(double value, int tapeIndex) {
    this.value = value;
    this.tapeIndex = tapeIndex;
  }
}
```

For each operation done on those numbers, an entry is recorded in the tape. The description of the tape entry is provided in Listing 4.5. The first information recorded is the type of basic operation performed. The list of those operations was provided above and in the code is described as an `enum`. The operations are elementary operations and operate on zero, one or two arguments. The tape index of the arguments are recorded in the `indexArg1` and `indexArg2` fields. If one or both arguments are irrelevant, the tape index is filled with -1. The next information is the value itself, this is the result of the operation. The *extra value* is some supplementary information required in some cases. This can be the multiplicative factor for the `MULTIPLICATION1`. It is also used for `MANUAL` entries as explained in a subsequent section.

Listing 4.5 Tape entry for Adjoint Algorithmic Differentiation by operator overloading

```
public class TapeEntryAad {
  private final OperationTypeAad operationType;
  private final int indexArg1;
  private final int indexArg2;
  private final double value;
  private final double extraValue;
  private double valueBar;
  ...
}
```

The last information is the value of the derivative of the final output with respect to the variable represented by the entry. It is called `valueBar` in the code to remind the notation used in the theoretical developments. This information is filled only in the backward sweep phase, which is called in this context the *interpretation* of the tape. That phase is described later.

The tape itself is described in Listing 4.6. It consists in a list of `TapeEntryAad`. One important information for the different values stored is the index of the corresponding operation. For that reason, when one entry is added to the tape, through the `addEntry` method, the index of the operation is returned to the main code.

Listing 4.6 Tape code

```java
public class TapeAad {
  private final List<TapeEntryAad> tapeList;
  private int size;

  public TapeAad() {
    this.tapeList = new ArrayList<TapeEntryAad>();
    size = 0;
  }
  public int size() {
    return size;
  }
  public TapeEntryAad getEntry(int index) {
    return tapeList.get(index);
  }
  public int addEntry(TapeEntryAad entry) {
    tapeList.add(entry);
    size++;
    return size - 1;
  }
}
```

Using those objects, the code for the example function described by Listing 2.1 in the previous chapter can be adapted easily. The code for the automatic Adjoint Algorithmic Differentiation version is given in Listing 4.7. With respect to the original code, the operators – like + – have been replaced by the method of the similar name and the mathematical functions – like exponential – have also been replaced by a method with similar name but using the `MathAad` method. For the rest the code looks very similar. Note that the tape need to be passed in each operation as each operation add an entry to the same tape and each operation create a new index.

Like for the standard AD case, the operations on the augmented doubles require a new math library. It is implemented in the `MathAad` class. An extract of that class is provided in Listing 4.8. The class consists of a list of static methods for the standard operations, like addition and multiplication and the basic functions, like exponential, sine, etc.

Listing 4.7 Simple function code by Automatic Adjoint Algorithmic Differentiation

```
static DoubleAad f_Aad_Automatic(DoubleAad[] a, TapeAad tape) {
  DoubleAad b1 = MathAad.plus(a[0],
    MathAad.exp(a[1], tape), tape);
  DoubleAad b2 = MathAad.plus(MathAad.sin(a[2], tape),
    MathAad.cos(a[3], tape), tape);
  DoubleAad b3 = MathAad.plus(MathAad.pow(a[1], 1.5d, tape),
    a[3], tape);
  DoubleAad b4 = MathAad.plus(MathAad.multipliedBy(
    MathAad.cos(b1, tape), b2, tape), b3, tape);
  return b4;
}
```

Listing 4.8 The mathematical library to be used with adjoint AD augmented doubles

```
public class MathAad {

  public static DoubleAad plus(DoubleAad d1, DoubleAad d2,
    TapeAad tape) {
    double valueOutput = d1.value() + d2.value();
    int index = tape.addEntry(new TapeEntryAad(
      OperationTypeAad.ADDITION, d1.tapeIndex(),
      d2.tapeIndex(), valueOutput));
    return new DoubleAad(valueOutput, index);
  }
  ...
}
```

The code presented in Listing 4.7 can be run on an example data. The resulting tape is presented in Table 4.1. The first column corresponds to the operation tape index or virtual address and the other columns to the information in the tape entries. The inputs used for the example are visible in the INPUT lines and are 0.0, 1.0, 2.0 and 3.0. The content presented is the tape after the interpretation which is described in later. Before the interpretation, the last column, with the derivatives values, is 0.0 for all entries.

The inputs are not computed from previous elements in the list and the argument indices are −1 for all of them. The first operation is the exponential which is applied to the value at index 1. The result, 2.7183, and the index are visible at line with index 4. The next line add two augmented doubles, the one at index 0 and the one at index 4, i.e. the second one is the exponential computed previously. The composition mechanism is starting here. The same process is performed for each line, up to the final addition at index 13. The output value is 4.0738.

The implementation proposed so far refers to the computation of the values themselves and the recording of the operations in the tape. All that make sense only if it is used later to compute the derivatives. The computation of the derivatives from the tape records is called the *tape interpretation*.

Table 4.1 Tape for the simple function described in Listing 2.1

Index	Operation	:	Arg 1	Arg 2	Value	Derivative
0:	INPUT	:	−1	−1	0.0	0.0331
1:	INPUT	:	−1	−1	1.0	1.5901
2:	INPUT	:	−1	−1	2.0	0.3794
3:	INPUT	:	−1	−1	3.0	1.1287
4:	EXP	:	1	−1	2.7183	0.0331
5:	ADDITION	:	0	4	2.7183	0.0331
6:	SIN	:	2	−1	0.9093	−0.9117
7:	COS	:	3	−1	−0.9900	−0.9117
8:	ADDITION	:	6	7	−0.0807	−0.9117
9:	POW1	:	1	−1	1.0	1.0
10:	ADDITION	:	9	3	4.0	1.0
11:	COS	:	5	−1	−0.9117	−0.0807
12:	MULTIPLICATION:		11	8	0.0736	1.0
13:	ADDITION	:	12	10	4.0738	1.0

The inputs are in the first four lines and are 0.0, 1.0, 2.0 and 3.0.

In our demonstration code, the interpretation is done in a method called `inter-pret`. The method is located in the class `TapeUtils`. The goal of that method is to implement the equivalent of the theoretical formula given by Eq. 2.4 to the practical case under consideration.

With respect to the theoretical formula, our task is facilitated as we work only on elementary operations that take at most two arguments. The operations are called g_k in the theoretical formula. One of the $\bar{b}[k]$ of the formula will play a role in at most two $\bar{b}[j]$ with $j > k$. Instead of working on the formula as written in the theoretical part and for one given j summing all the relevant k, we take one k at a time and see for which j it plays a role. Those j's are easy to obtain through the tape entries, those are given by the tape index of the operation inputs. The quantities $\frac{\partial}{\partial b_j} g_k$ still need to be computed. The formulas for those quantities, for elementary operation, can be found in any calculus book. In Listing 4.9 we have reproduced three of those operations. The addition of two augmented doubles, the addition of an augmented double with a primitive double and the sine of an augmented double. Once more, all the detailed formulas and cases can be found in the associated code.

The interpretation starts, as described in Table 2.3, by filling the derivative entry of the output with the value of the derivative of the output with respect to the output, this is 1.0. This step is done in line 3 of Listing 4.9. Then each entry is run through in reverse order. This is done using the loop defined in line 4.

The interpretation of the `SIN` operation is done in the following way. The sine takes only one argument, so only one derivative is impacted. The derivative variable for the entry referred by the index of the first argument will be impacted, and only that number. The index is retrieve in line 22. The derivative of the sine function applied to a double is the cosine applied to the same number – line 23. The value is recorded in the tape entry with index given by the same first argument. That cosine value is multiplied by the $\bar{b}[k]$, this is the derivative value for the entry

Listing 4.9 Tape interpreter for Adjoint Algorithmic Differentiation by operator overloading

```
1   public static void interpret(TapeAad tape) {
2     int nbEntries = tape.size();
3     tape.getEntry(nbEntries-1).addValueBar(1.0d);
4     for(int loope = nbEntries-1; loope>=0; loope--  ) {
5       TapeEntryAad entry = tape.getEntry(loope);
6       switch (entry.getOperationType()) {
7         case INPUT:
8           break;
9           ...
10        case ADDITION: // Addition of two DoubleAads.
11          tape.getEntry(entry.getIndexArg1())
12            .addValueBar(entry.getValueBar());
13          tape.getEntry(entry.getIndexArg2())
14            .addValueBar(entry.getValueBar());
15          break;
16        case ADDITION1: // Addition of a DoubleAad with a primitive double.
17          tape.getEntry(entry.getIndexArg1())
18            .addValueBar(entry.getValueBar());
19          break;
20          ...
21        case SIN: // Sine of a DoubleAad
22          tape.getEntry(entry.getIndexArg1()).addValueBar(
23            Math.cos(tape.getEntry(entry.getIndexArg1()).getValue())
24            * entry.getValueBar());
25          break;
26          ...
27        default:
28          break;
29      }
30    }
31  }
```

we are working with – line 24. By running through the tape in reverse order we cover all the recorded operations down to the input. When the loop is finished, all the required derivatives have been computed. The results can be collected easily. It suffice to go through the interpreted tape a collect all lines with an INPUT operation. In the code this is done by the extractDerivatives method of the class TapeUtils.

Up to now we have provided one implementation for automatic Standard AD and one for automatic adjoint AD and explained how to use the code for the very simple starter function described in Listing 2.1. We thus have now numerous versions of the derivative code for that simple function: four finite difference implementation, three standard AD – two manual and one automatic – and three adjoint AD – two manual and one automatic. From the versions discussed in previous chapter, we keep only three versions for further comparisons: the forward finite difference and the optimized manual standard AD and adjoint AD. To those versions we now add the automatic versions. The performance results are provided in Table 4.2.

Table 4.2 Computation time for the example function and four derivatives

Repetitions	Computation	Time	Ratio
100,000	Function	110	1.00
100,000	Function + FD forward	575	5.23
100,000	Function + SAD Optimized	195	1.77
100,000	Function + SAD Automatic	310	2.82
100,000	Function + AAD Optimized	165	1.50
100,000	Function + AAD Automatic	350	3.18
100,000	Function + AAD Automatic (no `interpret`)	260	2.36

Figures in milliseconds. Each repetition is with 5 data sets so that there is actually 500,000 repetitions.

For this very simple function, the manually crafted implementation perform significantly better than the automatic ones, both for forward and adjoint AD. For the standard mode, the derivatives add 0.77 to the function cost for the manual version and 1.82 for the automatic one. For the adjoint version, the figures are 0.50 and 2.18 respectively. In the case of the automatic adjoint version, we have also assessed the time to run the code using the augmented doubles and storing the tape but without running the interpretation code. This can be viewed as the extra cost to manipulate the data in an automatic way, but not doing the actual derivatives computation. Doing the actual derivative computation add only a cost of 0.82, which is not too far from the manual cost. From this very simple example one can see that the data manipulation with the augmented doubles is the part of the process that is adding the most to the computation time. The actual derivatives computations are still very efficient.

4.3   Automatic Algorithmic Differentiation Applied to Finance

In this section, we repeat the analysis done at the end of the previous section but for the functions related to finance analyzed in Chapter 3.

The first of those functions is the Black price function for options. The performance results are displayed in Table 4.3. For this simple function with only five inputs, the automatic AD does not perform as well as one would like. The recording of the tape takes a lot of time with respect to the simple operations involved. This can be seen in the last line where we performed the computation of the value using the augmented doubles, but without the interpretation of the tape, i.e. without actually computing the derivative. That recording on its own reduce the speed by a factor of six with respect to s straight implementation with primitive doubles.

The automatic AD version has still the advantage of the precision, but the speed advantage with respect to the single side (forward) finite difference implementation is lost. This is specific to the low number of inputs in our example. Nevertheless it appears to be a relatively standard result in implementations that the augmented doubles slow down the value computation by a factor between two to six and that the actual derivatives computation can be three to ten times the value computation time, against a maximum of three to four in theory and a factor of one to five in different manual implementation we analyzed.

Table 4.3 Computation time for the Black function and five derivatives

Repetitions	Computation	Time	Ratio
100,000	Function	75	1.00
100,000	Function + FD forward	525	7.50
100,000	Function + SAD	235	3.36
100,000	Function + SAD Automatic	575	8.21
100,000	Function + AAD Optimized	125	1.79
100,000	Function + AAD Automatic	760	10.13
100,000	Function + AAD Automatic (no `interpret`)	450	6.00

Figures in milliseconds. Each repetition is with 5 data sets and call/put so that there is actually 1,000,000 repetitions.

Table 4.4 Computation time for the SABR volatility function and seven derivatives

Repetitions	Computation	Time	Ratio
100,000	Function	75	1.00
100,000	Function + FD forward	630	8.40
100,000	Function + AAD Optimized	130	1.73
100,000	Function + AAD Automatic	860	11.47
100,000	Function + AAD Automatic (no `interpret`)	520	6.93

Figures in milliseconds. Each repetition is with 5 data sets so that there is actually 500,000 repetitions.

The next example is the SABR volatility function that was described in Listing 3.5. We implemented only the AAD version of the function and not the SAD one. The performance results are presented in Table 4.4. In this case also, the automatic version does not compete very well with the manual implementation and barely with the finite difference version in term of speed.

The last example in this section is the SABR with implied volatility option price. The code is the composition of the two previous methods. The performance results are presented in Table 4.5. We see the real use of the AD philosophy of applying derivatives to composition. Now there are also more inputs, in total the function has eight inputs. The automatic AAD provides results in a time roughly similar to the finite difference but with the advantage for precision. The function value and eight derivatives take roughly 12 value computation time. Note also that even for this relatively simple method, the AAD tape recording is of non-negligible length; in total the tape for SABR price has 81 entries.

4.4   Mixed Algorithmic Differentiation Implementations

By mixed Algorithmic Differentiation, we mean a mixture of manually written code and automatic operator overloading code. Some formulas that are used on a regular basis or are time consuming can be manually optimized while other that are less important performance-wise can be implemented as straight forward automatic AD.

Table 4.5 Computation time for the SABR option price function and eight derivatives

Repetitions	Computation	Time	Ratio
100,000	Function	240	1.00
100,000	Function + AAD	440	1.83
100,000	Function + AAD Automatic	2935	12.23
100,000	Function + AAD Automatic (no `interpret`)	1140	4.65

Figures in milliseconds. Each repetition is with 5 data sets and call/put so that there is actually 1,000,000 repetitions.

For this reason we have created in our automatic AD operation types, referenced in the enumeration `OperationTypeAad`, a semi-mysterious type called `MANUAL`. This is the mechanism by which the developer can introduce the result of derivatives computation in the automatic code. The tape entries with that type have the `extraValue` field populated. The extra-value is the derivative of the variable referenced by the entry with respect to one of the previous variables that were used to compute it. That previous variable does to need to be the immediately previous variable in the composition chain, it can be any previous variable. This allows the derivative code to jump several SAC lines. In practice for the examples we provide, we will directly jump from the inputs to the output of a method, without recording any of the method intermediary steps in the tape. The previous variable to which the derivative in the extra-value refers is indicated by the index in `indexArg1`. This way of bypassing the automatic AD while still keeping the automatic structure has two advantages with respect to a fully automated AD. The optimization done in the code by manual AD can be transferred to the automatic AD and there are less SAC lines recorded in the tape, so less memory requirements.

When it comes to the interpretation of the tape, the derivative of the argument variable is incremented by the extra-value multiplied by the derivative of the output with respect to the analyzed variable. In the notation of Eq. 2.4, this is $\bar{b}[\text{indexArg1}]+ = \bar{b}[k] \cdot \frac{\partial}{\partial b[\text{indexArg1}]} g_k$, where k is the index of the variable analyzed. This is the standard multiplication representing the composition. As we are interested by the first order derivative, we don't need to know really how the computation to obtain the variable, described by the function g_k, was performed. Only its first order derivative is important.

But with what we have described above, only one of the arguments of the skipped code is automatically interpreted for AD. A second variable could be added in the `indexArg2`. But the manual part result can a priori depends on more than two arguments, using the second index in that way would not solve the general case. Remember that the tape's entries have only two arguments because the elementary operation which are recorded by the automatic code have at most two arguments. The method we propose here to solve this problem without adding more arguments, is to have the second argument pointing to another version of the same variable where the second input variable derivative is represented in the extra-value. But that second version does not have the derivative of the output with respect to the analyzed variable, as recursively we have reached only the last version of that variable in the tape. It is then enough to copy that derivative of the analyzed variable to the `valueBar` field of that other copy to be able to continue the process. We create as many copies of the variable as number of input in the manual code. Each copy, but the first, pointing to the previous copy to continue the recursive process. The part of the `interpret` method which deal with automatic adjoint AD computation for the MANUAL type is represented in Listing 4.10.

Listing 4.10 The part of the tape `interpret` method related to the MANUAL operation

```
...
  switch (entry.getOperationType()) {
    case INPUT:
        ...
    case MANUAL:
        tape.getEntry(entry.getIndexArg1()).addValueBar(
          entry.getExtraValue() * entry.getValueBar());
        if (entry.getIndexArg2() != -1) {
          tape.getEntry(entry.getIndexArg2())
            .addValueBar(entry.getValueBar());
        }
        break;
    ...
```

With the previous explanation on the way to incorporate manual code in an automatic AD setting, we are ready to review some of our previous examples. In the first financial code we have analyzed, we looked at the Black formula. The manual code with automatic AD wrapping is available in the method `price_Aad_Automatic2` of the class `BlackFormula`. The code in that method is the same that in the `price_Aad_Optimized` for the main part. Only at the end, the storage of the manual computation is done in the relevant tape entries. From a signature point of view, the method looks like the automatic version. The last part of the relevant code is proposed in Listing 4.11.

The performance of this new implementation is compared to the one obtained earlier is displayed in Table 4.6. The performance is improved with respect to the full automatic version. The optimization done in the manual code still produces its effects. Nevertheless, the recording and manipulation of tapes is still significantly slower that dealing with primitive double.

The main philosophy of algorithmic differentiation is to use the composition efficiently. How does the manual/automatic duality transfers to composition. To show some possibilities we reimplemented the SABR with implied volatility option price with several mixed manual/automatic combination. The first one is an automatic implementation where both the SABR volatility and the Black price are computed by the manual version wrapped in the automatic language. This is the composition of the equivalent of two `Automatic2` codes as described above. The second implementation, called *Mixed M 1* uses the appearance of manual

Listing 4.11 The final part of the manual code with automatic wrapping for the Black formula

```
. . .
  int indexPrice0 = tape.addEntry(new TapeEntryAad(
    OperationTypeAad.MANUAL, forwardAad.tapeIndex(),
    price, inputBar[0]));
  int indexPrice1 = tape.addEntry(new TapeEntryAad(
    OperationTypeAad.MANUAL, volatilityAad.tapeIndex(),
    indexPrice0, price, inputBar[1]));
  int indexPrice2 = tape.addEntry(new TapeEntryAad(
    OperationTypeAad.MANUAL, numeraireAad.tapeIndex(),
    indexPrice1, price, inputBar[2]));
  int indexPrice3 = tape.addEntry(new TapeEntryAad(
    OperationTypeAad.MANUAL, strikeAad.tapeIndex(),
    indexPrice2, price, inputBar[3]));
  int indexPrice4 = tape.addEntry(new TapeEntryAad(
    OperationTypeAad.MANUAL, expiryAad.tapeIndex(),
    indexPrice3, price, inputBar[4]));
  return new DoubleAad(price, indexPrice4);
```

Table 4.6 Computation time for the Black function and five derivatives

Repetitions	Computation	Time	Ratio
100,000	Function	75	1.00
100,000	Function + AAD Optimized	125	1.79
100,000	Function + AAD Automatic	760	10.13
100,000	Function + AAD Automatic / Manual	530	7.07

Figures in milliseconds. Each repetition is with 5 data sets and call/put so that there is actually 1,000,000 repetitions.

AD but inside, the implied volatility is obtained by manual code and the Black formula by automatic code. The third implementation, called *Mixed A 1*, has the automatic appearance and inside, the volatility is automatic code and the Black formula is manual code. The fourth implementation, called *Mixed A 2*, has the automatic appearance and inside, the volatility is manual code and the Black formula is automatic code. There would be more combination possible, but it seems enough to have a view of the impact. The performance results are displayed in Table 4.7.

The main conclusion from this section is that the manual and automatic algorithmic differentiation can be used in combination. This is the real philosophy of AD as I see it: use the *efficiency of composition* to achieve the *best result*. There is no a priori superiority of one approach with respect to another. The secondary conclusion is that automatic AD is less efficient that expertly crafted AD and that the ratio between the two can be larger than five. We will come back to the advantages of expertly crafted AD code in the next chapter.

Table 4.7 Computation time for the SABR price function and eight derivatives

Repetitions	Computation	Time	Ratio
100,000	Function	240	1.00
100,000	Function + AAD Optimized	440	1.83
100,000	Function + AAD Automatic	2935	12.23
100,000	Function + AAD Automatic 2	980	4.08
100,000	Function + AAD Mixed M 1	1230	5.13
100,000	Function + AAD Mixed A 1	2480	10.33
100,000	Function + AAD Mixed A 2	1340	5.58

Figures in milliseconds. Each repetition is with 5 data sets and call/put so that there is actually 1,000,000 repetitions.

Bibliography

Naumann, U. (2012). *The Art of Differentiating Computer Programs. An Introduction to Algorithmic Differentiation.* Philadelphis:SIAM.

Naumann, U. and du Toit, J. (2014). Adjoint algorithmic differentiation tool support for typical numerical patterns in computational finance. Technical report, RWTH Aachen University.

Chapter 5
Derivatives to Non-inputs and Non-derivatives to Inputs

You didn't know you wanted it – You get what you have not asked for – Volatility can stick.

The title of this chapter may appear a little bit cryptic. The main advertised goal of Algorithmic Differentiation (AD) is to compute in an efficient way the derivatives of functions with respect to their inputs. Each part of this chapter's title may appear in contraction with that general goal.

Nevertheless, to my opinion, the content of this chapter is highly relevant for this book in the context of application of AD to finance. Instead of reading it as a negation of the book's main goal, the content of this chapter should be viewed as an extension of it. A less cryptic, but significantly longer and less fun title could have been: "computing derivatives with respect to financially meaningful numbers that are not direct inputs to the global computation and computing numbers closely related to derivatives but not matching exactly the theoretical definition of derivative."

The techniques described below are in general not available to automatic AD. They are producing results relevant for a subject matter expert but that do not appear directly in the data structure. Automating the interpretation of the financial meaning of code is probably beyond the current reach of programming and will need to wait for more developments in artificial intelligence. We still need to trust human experts to extract as much insight as possible from the existing developments.

5.1 Derivatives with Respect to Non-inputs

As a first example, we look at the question of the vega hedging of caps and floors. Suppose that we use a description of the cap price through normal[1] implied volatilities. In our example, the so-called "smile" of normal volatilities is described for this purpose by a set of fixed strikes volatilities and an interpolation mechanism is

---

[1] We use normal/Bachelier (1900) volatilities and not log-normal/Black (1976) volatilities to avoid the problem with negative strikes and rates.

© The Author(s) 2017
M. Henrard, *Algorithmic Differentiation in Finance Explained*,
Financial Engineering Explained, DOI 10.1007/978-3-319-53979-9_5

used to obtain the volatility between them. For the moment we ignore the expiry dimension.

A direct implementation of algorithmic differentiation will provide the sensitivity of the price to the different strike nodes describing the smile, this is to a fixed set of strikes. This may not be what the financial user is interested in. Even if the smile is described using fixed strikes node strikes, those strikes may not be fundamentally different from others strikes. A good complete set of financial information provided to the risk manager will contains on one side the sensitivity to the interpolated volatility used in the price computation and on the other side the sensitivity to each individual node used to describe the smile. The volatility interpolated between the nodes is not one of the original input data provided but an intermediary number which is part of the computation. The business user may be interested by information on derivatives with respect to intermediary values. The architecture of the system should allow him to request and obtain those numbers. Moreover the computation time for providing one or both of the information described above should not be fundamentally different. The sensitivity to an intermediary value should be available without restarting the full computation.

In the swap market, swaps with yearly tenors are liquid and the building blocks of the market. The interest rate curves are calibrated with those data points and the macro-hedging of a portfolio is usually done with the same instruments. The inputs use in the computer implementation are matching the information the trader uses. For example trading a 5-year, 7-month and 10-day swap to hedge some interest rate level is cumbersome and attract a larger bid/offer due to illiquidity. In this context, computing the sensitivity with respect to the yearly swap rates for an interest rate portfolio makes sense, both from a financial and a technological point of view.

We have already described a similar requirement for interest rate curve in Section 3.3. The information provided to the end-user for interest rate curves should include the point sensitivity, the parameter sensitivity and the market quote sensitivity or in a more generic language, the sensitivity to each individual point used in the computation, to the model parameters and to the market data used to calibrate the model.

A natural number computed for interest rate books through AD is the sensitivity of the book to the different market quotes used to calibrate the curves. This is a Risk Measurement number. To achieve Risk Management, the risk manager has to convert this number into an action. The question is: *Which amount of the financial product represented by the market quote should I buy/sell to eliminate the risk described by the sensitivity?*

From the market quote sensitivities $\partial PV / \partial q_k$ computed in the risk measurement step, one would like to establish the notional of each instrument that should be traded. First note that the present value PV_k of each instrument with unit notional k used in the curve calibration, is such that

$$\frac{\partial PV_k}{\partial q_l} = 0 \quad \text{for} \quad k \neq l.$$

Each instrument has a present value of zero if its quote is the one used in the calibration step. Changing the other quotes does not affect that value. The value of a given instrument used in the curve calibration is not sensitive to the change of quotes associated to the other instruments. We have only one non-zero number in those sensitivities for each node k; we denote it

$$T_k = \frac{\partial PV_k}{\partial q_k}$$

for a calibrating instrument with unit notional.

This number is easy to compute, it suffices to apply the principles described in the previous section for the present value sensitivity to each of the n instruments in turn, obviously using AD. That step can be done once at curve calibration time and stored to be used as described below as many time as necessary.

Suppose that the sensitivity of a portfolio to the different market quotes is given by

$$s_k = \frac{\partial PV}{\partial q_k}.$$

To obtain a portfolio with first order sensitivity equal to zero for all buckets, it is enough to trade a notional

$$N_k = -\frac{s_k}{T_k}$$

in each of the calibration instruments.

Adding those trades to the portfolio we obtain a total sensitivity to the market quote q_k of

$$s_k + N_k \cdot T_k = s_k - \frac{s_k}{T_k} \cdot T_k = 0.$$

As promised we obtain the perfect (local) hedging of the portfolio with the instruments used to calibrate the curve at the cost of one multiplication for each instrument in the calibration set. The multiplication factor $-1/T_k$ is computed as a by-product of the AD implementation. It is related to the derivative of the PV of an intermediary instrument computed during the curve calibration process. It is useful for the risk manager to be able to request that quantity to be available for further processing. The quantity $-1/T_k$ is not strictly speaking the derivative with respect to one of the input, but the derivative of the theoretical quote when the present value of the underlying instrument changes. It not a quantity automatically computed, but a portfolio manager or trader will certainly find it a very useful quantity to have at his disposal.

5.2   Non-derivatives with Respect to Inputs

In most cases, the inputs to valuation are market data, this is market quotes of financial instruments. For some instruments, like bonds and futures, the quote may be a price, for others, like FRA, IRS and inflation swaps, the quote may be a rate and for other, like basis swaps or Forex swaps, it may be a spread. For some, the quoting mechanism can be indirect, like implied volatility for options or yield for bonds; a standard or conventional formula needs to be used to obtain the actual term sheet of the trade.

Using the conventional formula is a mechanism to facilitate the communication between the market participant, it is not a religion nor an indication that the user believes that the intuition behind the formula is correct. It is not because someone quotes the Black or Bachelier formula implied volatility for an option, that he necessarily uses the log-normal or normal dynamic implicit from the formula to risk manage his option book. Even if we restrict ourselves to a Black formula with an implied volatility smile or surface, the description of the current smile, which is what can be observed in the market, should not be read as a estimation on how it will evolve in the future. Someone may describe the smile of interest rate cap/floors in term of implied volatility by strike, this does not implies that he believes that the best prediction for the future is that the implied volatility for a given strike will be constant. Predicting that the implied volatility for a given strike is stable when the underlying market evolves is called a *sticky strike* dynamic for obvious reasons. The fact that cap/floor market conventionally quote volatility by strike does not mean that all cap/floor market participants use the sticky strike approach for risk management purposes.

One of the direct applications of Algorithmic Differentiation in finance is the computation of greeks. The algorithmic differentiation technique itself is the art of computing derivatives of the output with respect to the input. There is a semantic slide between the two previous sentences. Greeks and derivatives are almost synonyms in finance, except in this section where they are almost antonyms. The greeks we are interested in may be different from the direct derivatives of the formula, they may be a combination of different derivatives to match a specific intuition or derivatives with respect to a number which is not explicitly a market quote but that can be implied from them.

In the martingale approach to pricing, the choice of model is almost synonymous of the choice of delta with respect to the underlying. The risk manager should have the choice of selecting the data that to his opinion provides the best representation of the current market and, separately, he should have the choice to decide of the modeling approach that represents his best view of the market future evolution. It is the job of the developer to allow those choices to be made by the end user. Not only the straight "derivative with respect to the input" should be available but also any sensible modification thereof as imagined by the risk manager. The terms "sensible" and "imagined" can be somewhat contradictory, this is why the dialogue between the two characters above is important.

Note that in the *Fundamental Review of Trading Book*, the Basel Committee on Banking Supervision (2014)[2] (BCBS) prescribes that the computation of

---

[2] http://www.bis.org/bcbs/

sensitivities for regulatory computation of capital related to the trading book should be done in a *sticky delta* approach – that we describe below. Even if I personally disagree with that recommendation, I agree with the fact that a risk manager should be able to choose his preferred way, whatever the market conventions are.

5.2.1 Sticky to Something – The Problem

In this section, we take a concrete example of the above abstract discussion and describe how to implement it in practice. In this case, we suppose we have a financial option priced by a Black formula. This can be an equity, forex or interest rate option. Obviously for interest rate options, the Black/log-normal model may not be the best choice as rates and strikes can be negative. If you are working in interest rate, you can replace Black by Bachelier everywhere in this section; all the results remain valid with very minor adjustments. It is also valid for any base formula depending of similar parameters: forward, strike and a unique model parameter.

We consider an option with strike K and time to expiry θ. The Black formula, already described in Eq. 3.1 is given, ignoring the numeraire which is not important for our discussion, by

$$\text{Black}(K, \theta, F_0, \sigma_0) = \omega \left(F_0 N(\omega d_+) - K N(\omega d_-) \right)$$

where $\omega = 1$ for a call, $\omega = -1$ for a put and

$$d_{\pm}(K, \theta, F_0, \sigma_0) = \frac{\ln\left(\dfrac{F_0}{K}\right) \pm \frac{1}{2}\sigma_0^2 \theta}{\sigma_0 \sqrt{\theta}}. \tag{5.1}$$

The base present value formula is

$$\text{PV}_0 = \text{PV}(K, \theta, F_0) = \text{Black}(K, \theta, F_0, \sigma_0) \tag{5.2}$$

with F_0 the underlying current forward price and σ_0 the implied volatility for that option. By implied volatility we mean the usual "wrong parameter in the wrong formula to obtain the correct price." By saying that, we implicitly mean that we do not trust the model that leads to the formula but we use the formula as a convenient mean to store information about the market through the implied volatility. We can observe the same market, not only for the strike price we are interested in, but for other strikes as well. We can obtain the current implied volatility for all strikes, called the *market smile*: $\sigma = \sigma^{\text{Mkt}}(K)$. For different strikes, the market prices are given by

$$\text{PV}(K, \theta, F_0) = \text{Black}(K, \theta, F_0, \sigma^{\text{Mkt}}(K)) \tag{5.3}$$

We are interest in the derivative of the price with respect to the forward price or rate. We cannot differentiate directly the PV function from Eq. 5.2. The formula gives the correct price for the current forward rate F_0 but there is no indication that the same formula would be correct if the underlying value was to change. The notation in the above formula may let us think that the volatility is simply strike dependent and we can differentiate Eq. 5.3. This is not the case either. The volatility shows only a strike dependency because of our ignorance, we don't know what happens for a different market F_0. The present value formula should actually be written as

$$\text{Black}(K, \theta, F, \sigma(K, \theta, F, \ldots)). \tag{5.4}$$

But we know only a small part of the volatility function, we know it only for one value of F – the current value F_0 – and we want to differentiate with respect to that same F. In other words, we are looking for the derivative in a direction where we know almost nothing. We want a short movie about the main character – the present value – but we only have a still picture of the opening sequence of the movie.

It means that we have to add some hypothesis to the framework to obtain what we are looking for. One potential way to select those hypothesis is to suppose a *sticky something* behavior for the smile. The standard "something" that can be sticky are *strike*, *moneyness* or *delta*.

Before going to the details of what we mean by *sticky*, we describe the other terms. The strike term does not need further explanation. The moneyness can be express as a *simple moneyness*, which is the difference between the strike and the forward $(K - F)$ or as a *log-moneyness*, which is the logarithm of the ratio of strike and forward $(\log(K/F))$. The delta is the Black formula delta given by $N(d_+)$ with d_+ described in Equation (5.1).

What does it mean that the volatility description *sticks* to one of those quantities? It means that we associate the volatility for a new market value $F \neq F_0$ by using the current market smile σ^{Mkt} with that quantity constant. Let $f(K, F)$ be the quantity that is preserved.

For the cases mentioned above, those quantities are

Strike: $f(K, F) = K$

Simple moneyness: $f(K, F) = K - F$

Log-moneyness: $f(K, F) = \log(K/F)$

Delta: $f(K, F, \sigma) = N(d_+(K, F, \sigma))$

The delta case is more involved as it refers to the volatility and will be treated separately later.

For a change in the underlying market to a different forward $F \neq F_0$, the new volatility to be used to price the option with strike K is

$$\sigma = \sigma^{\text{Mkt}}(L) \quad \text{where } L \text{ is such that} \quad f(K, F) = f(L, F_0). \tag{5.5}$$

The volatility to use is given by the market volatility at an implied strike L. This implied strike is obtained by using a conservation law.

For the sticky strike case, the solution is easy, $F(K, F) = K$, $L = K$ and

$$\sigma = \sigma^{\text{Mkt}}(K).$$

This is the simplest case; the implied volatility for a given option with a given strike does not change with the change of the underlying market.

For the sticky moneyness case, the solution of $K - F = L - F_0$ is given by $L = K - F + F_0$ and the volatility to be used is

$$\sigma = \sigma^{\text{Mkt}}(K - F + F_0).$$

The new volatility is obtained by a simple translation.

By the sticky extension, we have an extension of the implied volatility to all forward F and we can write

$$PV(K, F) = \text{Black}(K, F, \sigma)$$

with σ satisfying Eq. 5.5.

5.2.2   Sticky Something, Volatility Independent

In this section, we suppose that the function f defining the stickiness depends on F and K only and not on σ.

Remember, what we are trying to compute is the sensitivity of the price to the change of the underlying forward price or rate. We are looking for the derivative with respect to F, the third variable in the present value function,

$$D_3 PV(K, \theta, F_0) = D_3 \text{Black}(K, \theta, F_0, \sigma_0) + D_4 \text{Black}(K, \theta, F_0, \sigma_0) D_3 \sigma(K, \theta, F, \ldots).$$

In the above formula, most of the pieces are known. The derivative of the Black formula with respect to the forward is called the (forward) *delta* and the derivative with respect to the volatility is called the *vega*. The formula is thus

$$D_3 PV(K, \theta, F_0) = \Delta(K, \theta, F_0, \sigma_0) + \text{Vega}(K, \theta, F_0, \sigma_0) D_3 \sigma(K, \theta, F, \ldots).$$

The only piece we still have to provide is the $D_3 \sigma$. For the two special cases of the sticky strike and sticky simple moneyness, the formula of the volatility in term of forward is explicit and we could compute the derivative directly.

In a more general setting, we can still obtain the result by using the implicit function theorem which is detailed in Appendix A.3. The implicit result is given by Equation (5.5). We have a point with a root of the equation: $f(K, F_0) - f(K, F_0) = 0$ and we would like to find an implicit function g which solved the equation

$f(K, F) - f(g(F), F_0) = 0$ for points around F_0. With that notation, the volatility is $\sigma(K, \theta, F, \ldots) = \sigma^{\text{Mkt}}(g(F))$. We want to compute $D_3\sigma$, which is possible by the implicit function theorem. Note that, by definition, $g(F_0) = K$.

By the implicit function theorem, the derivative of g is given by

$$D_1 g(F_0) = \frac{D_2 f(K, F_0)}{D_1 f(K, F_0)}.$$

We have not mentioned yet the proof of the existence of g; by the same implicit function theorem, the function g exists around F_0 if $D_1 f(K, F_0) \neq 0$.

If this is the case, we have as final solution

$$D_3\text{PV}(K, \theta, F_0) = \Delta(K, \theta, F_0, \sigma_0) + \text{Vega}(K, \theta, F_0, \sigma_0)D\sigma^{\text{Mkt}}(K)\frac{D_2 f(K, F_0)}{D_1 f(K, F_0)}.$$

Obviously this requires that σ^{Mkt} is differentiable. In particular a simple linear interpolation of market volatilities would not provide very good results.

In the *sticky strike* case, we have $F(K, F) = K$ and $D_2 f(K, F_0) = 0$. So there is no vega part in the formula and

$$\Delta_{\text{StickyStrike}} = D_3\text{PV}(K, \theta, F_0) = \Delta(K, \theta, F_0, \sigma_0).$$

In the *sticky simple moneyness* case, we have $F(K, F) = K - F$, $D_2 f(K, F_0) = -1$ and $D_K f(K, F_0) = 1$. So the final solution is

$$\begin{aligned}\Delta_{\text{StickyMoneyness}} &= D_3\text{PV}(K, \theta, F_0) \\ &= \Delta(K, \theta, F_0, \sigma_0) - \text{Vega}(K, \theta, F_0, \sigma_0)D_1\sigma^{\text{Mkt}}(K). \end{aligned} \quad (5.6)$$

In the *sticky log-moneyness* case, we have $F(K, F) = \log(K/F)$ and $D_2 f(K, F_0) = -1/F_0$ and $D_1 f(K, F_0) = 1/K$. So the final solution is

$$\begin{aligned}\Delta_{\text{StickyLogMoneyness}} &= D_3\text{PV}(K, \theta, F_0) \\ &= \Delta(K, \theta, F_0, \sigma_0) - \text{Vega}(K, \theta, F_0, \sigma_0)D_1\sigma^{\text{Mkt}}(K)\frac{K}{F_0}. \end{aligned} \quad (5.7)$$

5.2.3 Sticky Something, Volatility Dependent

The sticky delta case is a slightly more complex problem. The reason is that the constraint is now of the form $f(K, F, \sigma) = f(L, F_0, \sigma^{\text{Mkt}}(L))$. The constraint f is on the delta and the delta itself depends on the σ. We cannot solve the two problems consecutively, we have to solve them simultaneously.

The problem is now

$$\sigma = \sigma^{\text{Mkt}}(L)$$
$$f(K, F, \sigma) = f(L, F_0, \sigma^{\text{Mkt}}(L))$$

where f is given by d_+ from Eq. 5.1. It can be written as a systems of equations, considering K and F_0 as constants, as

$$h(L, F, \sigma) = 0$$

We now have a system of two equations with three unknowns L, F and σ. The goal is to obtain (implicitly) $\sigma(F)$ and (explicitly) $D\sigma(F_0)$. The partial derivatives we need are given by

$$D_{(1,3)}h(K, F_0, \sigma_0) = \begin{pmatrix} -D_1\sigma^{\text{Mkt}}(K) & 1 \\ -D_1f(K, F_0, \sigma_0) - D_3f(K, F_0, \sigma_0)D_1\sigma^{\text{Mkt}}(L) & D_3f(K, F_0, \sigma_0) \end{pmatrix}$$

$$D_2h(K, F_0, \sigma_0) = \begin{pmatrix} 0 \\ D_2f(K, F_0, \sigma_0) \end{pmatrix}$$

And the inverse to be used is

$$\left(D_{(1,3)}h(K, F_0, \sigma_0)\right)^{-1} =$$

$$\frac{1}{D_1f(K, F_0, \sigma_0)} \begin{pmatrix} D_3f(K, F_0, \sigma_0) & -1 \\ D_1f(K, F_0, \sigma_0) + D_3f(K, F_0, \sigma_0)D_1\sigma^{\text{Mkt}} & -D_1\sigma^{\text{Mkt}}(K) \end{pmatrix}$$

After easy algebraic manipulations, we have

$$D_F\begin{pmatrix} L \\ \sigma \end{pmatrix} = -\frac{1}{D_1f(K, F_0, \sigma_0)} \begin{pmatrix} -D_2f(K, F_0, \sigma_0) \\ -D_2f(K, F_0, \sigma_0)D_1\sigma^{\text{Mkt}}(K) \end{pmatrix}$$

The final result for the case of the sticky delta is

$$D_F\text{PV}(F_0, K) = \Delta(F_0, \sigma_0) + \text{Vega}(F_0, \sigma_0)D_1\sigma^{\text{Mkt}}(K)\frac{D_2d_+(K, F_0, \sigma_0)}{D_1d_+(K, F_0, \sigma_0)}$$

with

$$D_1d_+(K, F_0, \sigma_0) = -\frac{1}{\sigma_0\sqrt{\theta}}\frac{1}{K}$$

$$D_2d_+(K, F_0, \sigma_0) = \frac{1}{\sigma_0\sqrt{\theta}}\frac{1}{F_0}.$$

This leads to a final result

$$\Delta_{\text{StickyDelta}} = D_3\text{PV}(K, \theta, F_0) = \Delta(K, \theta, F_0, \sigma_0) - \text{Vega}(K, \theta, F_0, \sigma_0)D_1\sigma^{\text{Mkt}}(K)\frac{K}{F_0}.$$

Note that the *sticky delta* risk is locally the same as the *sticky log-moneyness* case.

In this section we have constructed different deltas, the original formula Delta, which is equivalent to the sticky strike Delta but also the sticky moneyness Delta, the sticky log-moneyness Delta and the sticky delta Delta. All those numbers have been obtained from the same set of market data, i.e. the non-dynamic market smile $\sigma^{\mathrm{Mkt}}(K)$. There is no need to manipulate the original data to obtain the different results. An implementation of pricing through Black formula with smile should provide the three different deltas as standard output. Then it is up to the risk manager to use one or all of them, or even to request a different one. The risk manager should not be required to transform the input data to be able to obtain the risk measure he would like to see; the system should provide this feature implicitly. The non-derivatives to the input can be computed directly in the library by using simple manipulations of the results provided by the AD framework. Remember, Algorithmic Differentiation is not a mere technique, it is really a *philosophy of implementation*.

Bibliography

Bachelier, L. (1900). *Théorie de la Spéculation*. PhD thesis, Ecole Normale Supérieure.

Basel Committee on Banking Supervision. (2014). Fundamental review of the trading book: outstanding issues. Consultative document, Basel Committee on Banking Supervision.

Black, F. (1976). The pricing of commodity contracts. *Journal of Financial Economics*, 3(1–2):167–179.

Chapter 6
Calibration

What we need is implicit – The fast lane to calibration.

The term *calibration* used in the chapter's title is the term used in finance for root finding or finding an optimal in the least-square sense. The type of problem encountered can be summarized in the following way. A certain number of financial market information, called *market quotes*, are available. This is for example the par rates of swaps, the prices of futures, the yields of bonds, or the prices of options. We want to use a specific model to compute the present value of other instruments, similar to the original ones, or to compute the risks associated to them. The model is based on a certain number of parameters. Those parameters can be the zero-coupon rates of a curve, the Black implied volatilities, or the volatility parameters of a Libor Market Model.

From the model, one can compute the *theoretical quotes* that would be prevalent if the model was correct. Those theoretical quotes depend on the model parameters. The calibration consists in searching for the parameters of the model which are such that the theoretical quotes are equal to or as close as possible to the market quotes. When the theoretical quotes are equal to the market quotes, we say that we have an *exact calibration*; when the theoretical quotes are not equal to the market quote and found through a least-square-like approach, we say that we have an *at-best calibration*. Usually in the exact calibration we have the same number of model parameters as the number of market quotes targeted. In the at-best calibration, there are more market quotes than parameters in the model.

6.1 Interest Rate Curves Calibration

In the approach we present in this section, we deal only with exact calibration for interest rate curves in the multi-curve framework. We suppose that we have a certain number of instruments in the calibration basket and the same number of parameters in the curves. The parameters are split among different curves. We suppose also that the multi-dimensional root finding problem defined by the constraints on each instrument and the curve parameter has one non-singular solution.

© The Author(s) 2017

M. Henrard, *Algorithmic Differentiation in Finance Explained*,
Financial Engineering Explained, DOI 10.1007/978-3-319-53979-9_6

It is quite difficult to describe financial conditions that lead to those mathematical hypothesis without prescribing specific curve descriptions and instrument restrictions. We would like to leave the approach as general as possible, so we refrain going in that direction.

We have n instruments that are used in the calibration and we index them by k ($1 \leq k \leq n$). Let $p = (p_l)_{1 \leq l \leq n}$ denote the parameters of the curves and $S = (S_k)_{1 \leq k \leq n}$ the functions for which we have to find a root for the k-th instrument. This is in general the market quote par spread or a similar function. The market quotes themselves are denotes $q = (q_k)_{1 \leq l \leq n}$. See Henrard (2014b) for more details on the curve descriptions and the functions to solve. The problem to solve is

$$S_k(p) = 0 \quad (1 \leq k \leq n). \tag{6.1}$$

In general, the problem is non-linear and strongly coupled. The solution has to be obtained globally for all parameters p as one computation block, it is not possible to bootstrap the different equations.

In some special cases, the problem can be divided in sub-problems, easier to solve numerically. The most extreme simplification is when the problem can be divided into one dimensional sub-problems. This is the case of the bootstrapping approach, where the interpolation scheme is local and the curves are not entangled. This is an extreme case, quite non-realistic in practice and we do not analyze it here.

A more frequent numerical simplification is when some curves are not fully entangled and the problem can be solved inductively. First calibrating some curves and then using the resulting curves as an input to calibrating the next curves. In the sequel we call *curve units* the set of curves calibrated simultaneously in one unique root-finding process. The complete set of all the related curves is called a *group*.

The standard application of this is when the curves are build one by one. More involved examples are when units of curves are first calibrated and then another unit of several curves is calibrated based on the previous units. Let u be the number of units and n_i the total number of parameters in the first i units. The steps in the induction will look like:

$$
\begin{aligned}
S_k((p_l)_{0 < l \leq n_1}) &= 0 \quad (0 < k \leq n_1) \\
S_k((p_l)_{0 < l \leq n_1}, (p_l)_{n_1 < l \leq n_2}) &= 0 \quad (n_1 < k \leq n_2) \\
&\vdots \\
S_k((p_l)_{0 < l \leq n_{u-1}}, (p_l)_{n_{u-1} < l \leq n}) &= 0 \quad (n_{u-1} < k \leq n)
\end{aligned}
\tag{6.2}
$$

Each vector equation represents one unit. In the function S_k above we represent only the parameters that impact the equalities.

Solving the above system of equations, this is finding the parameters $(p_l)_{0 < l \leq n}$, should not be the only output of the calibration exercise. The output of the procedure should facilitate the computation of derivatives related to other functions. The calibration function is only one step in the general AD implementation for interest rate.

For example the calibrated curves are used to compute the present value of instruments – denoted PV. What we are also interested in are the sensitivities or first order derivatives of the present value to the market quotes q_k used in the curve building procedure, i.e. we are also interested in

$$\frac{\partial \text{PV}}{\partial q_k}.$$

Those numbers are often called the *bucketed deltas*, *partial PV01* or *key rate duration* of the instruments.

Using the calibrated curves, it is possible to compute the derivatives of the present value to the curves parameters p. Algorithmic differentiation generally helps for that also. The derivative with respect to the market quotes can be obtained through the usual derivative of composition by

$$\frac{\partial \text{PV}}{\partial q_k} = \sum_{l=1}^{n} \frac{\partial \text{PV}}{\partial p_l} \frac{\partial p_l}{\partial q_k}.$$

This is the usual composition used in AD, first calibrate the curves, that is compute the function $p(q)$, and then compute the present value of the other instruments. What we have not described above is how to compute in practice the matrix

$$D_q p = \left(\frac{\partial p_l}{\partial q_k} \right)_{1 \leq l \leq n, 1 \leq k \leq n}.$$

That matrix is often called *inverse Jacobian matrix* or *inverse transition matrix*. Obviously you could implement AD for the full root-finding procedure and par spread computation and obtain the result that way. What we show below is that this is not required, it is sufficient to have the AD implemented for the par spread part.

The functions we suggest to use for the root-finding procedure in the calibration are the par market quotes spreads. They are the additive spreads to the market quotes for which the present value of the instrument is equal to zero, this is for q^{Mkt} the market quotes and $\tilde{q}$ the computed quotes from the curves,

$$S(p, q^{\text{Mkt}}) = \tilde{q}(p) - q^{\text{Mkt}}. \tag{6.3}$$

We rewrite Eq. 6.1 to indicate explicitly the market quote dependency. The calibrated parameters p_0 are such that

$$S(p_0, q^{\text{Mkt}}) = 0.$$

The market quotes q^{Mkt} are fixed numbers and are not part of the parameters to change to find the root. Provided that some regularity conditions are satisfied around the solution (p_0, q^{Mkt}), one can apply the *implicit function theorem* described in Appendix A.3. There is a function $p : \mathbb{R}^n \rightarrow \mathbb{R}^n; q \mapsto p(q)$ defined in a neighborhood of q^{Mkt} such that

$$S(p(q), q) = 0$$

for all q in the neighborhood. Using the derivative part of the implicit function the-orem, the derivatives of the calibrated parameters with respect to the quotes q at the market quote q^{Mkt} can be obtained by

$$D_q p(q^{\text{Mkt}}) = -(D_p S)^{-1}(p_0, q^{\text{Mkt}}) D_q S(p_0, q^{\text{Mkt}}) = (D_p S)^{-1}(p_0, q^{\text{Mkt}}).$$

The last equality is obtained using the fact that the derivative of S given in Eq. 6.3 to the quote q is -1.

When the root-finding problem is solved with a Newton-like algorithm, the same type of matrix is computed internally in the solver. The Jacobian matrix is used to find the direction of the next best guess. Computing the matrix for later use does not require developments beyond those already done for solving efficiently the root-finding algorithm. In what follows, we will call the *Jacobian* or *transition* matrix the matrix $D_p S$ and the *inverse Jacobian* the matrix $D_q p = \left(D_p S \right)^{-1}$.

We now describe a way to improve further the efficiency of the computation of the above matrix. To obtain the transition matrices between the market quotes and the curve parameters, one can compute directly the huge matrix with all the market quotes as input and all the curve parameters as the output of a very large root-solving problem.

This is possible, but if you work in a multi-currency and multi-curve framework with collateral – see (Henrard 2014b, Section 8.3) – with many currencies, many collateral and many ways to combine them, one has to deal easily with 10 curves with 20 parameters each, this is a total of 200 parameters and market quotes. Solving a 200×200 non-linear system of equations and computing the associated Jacobian matrix is probably not the most efficient way to achieve the result. As described above in the system of equations (6.2), the calibration can be done in an inductive way for most of the curve settings. The last curves calibrated depend on all the previous curves, but at each step only a reduced system has to be solved. Usually the sub-system, that we call unit, contains only one to three curves and not ten or more like the full system.

Suppose that the curves are obtained as a multidimensional root-finding process in an inductive way. The goal is to obtain the calibrated curves $(p_l)_{l \in (0,n]}$ from the market rates $(q_k)_{k \in (0,n]}$ but also the generalized transition matrix

$$\left(D_{q_k} p_l \right)_{l \in (0,n], k \in (0,n]}.$$

Due to the way the curves are built, we know that a good part of the matrix is full of zeros; the non-zero part, which we are interested in, is made of the sub-matrices

$$\left(D_{q_k} p_l \right)_{l \in (n_{i-1}, n_i], k \in (0, n_i]}.$$

The equations solved to obtain the parameters are at the i-th step:

$$S(n_{i-1}, n_i](p(0, n_{i-1}], p(n_{i-1}, n_i]) = 0 \tag{6.4}$$

where the notation $x(i, j]$ represents the vector with values $(x_k)_{i<k\leq j}$. We have split the parameter vector in two parts: the parameters that have already been calibrated $((0, n_{i-1}])$ and the parameters we calibrate in this step $(n_{i-1}, n_i]$.

Suppose that at each step, the previous step transition matrices $D_{q_k}p_l$ are available for $l \in (n_{i-2}, n_{i-1}]$ and $k \in (0, n_{i-1}]$. We can compute $D_{p_l}q_k = D_{p_l}S_k$ for $l, k \in (n_{i-1}, n_i]$ directly as described previously. Moreover

$$\left(D_{q_k}p_l\right)_{l,k\in(n_{i-1}, n_i]} = \left(\left(D_{p_l}q_k\right)_{l,k\in(n_{i-1}, n_i]}\right)^{-1}.$$

So we have the derivative of the new set of parameters $(p_l)_{(n_{i-1}, n_i]}$ with respect to the new market rates $(q_l)_{(n_{i-1}, n_i]}$. For the previous market rates, we use composition. Suppose that we have $D_{p_k}p_l$ for $l \in [n_{i-1}, n_i)$ and $k \in [0, n_{i-1})$. Then

$$\left(D_{q_k}p_l\right)_{l\in(n_{i-1}, n_i], k\in(0, n_{i-1}]} = \left(D_{p_m}p_l\right)_{l\in(n_{i-1}, n_i], m\in(0, n_{i-1}]} \cdot \left(D_{q_k}p_m\right)_{m\in(0, n_{i-1}], k\in(0, n_{i-1}]}$$

The second factor is provided by the previous steps, we still have to obtain the first factor. Using the implicit function theorem for the Equation (6.4), we have

$$\left(D_{p_m}p_l\right)_{l\in(n_{i-1}, n_i], m\in(0, n_{i-1}]} = -\left(\left(D_{p_l}S_j\right)_{j\in(n_{i-1}, n_i], l\in(n_{i-1}, n_i]}\right)^{-1} \left(D_{p_m}S_j\right)_{j\in(n_{i-1}, n_i], m\in(0, n_{i-1}]}.$$

We now have all the required results to keep track of the full transition matrix.

In the following example we use an implementation[1] of the above construction. We build four curves, all of them for instruments collateralized at USD Fed Fund: USD Fed Fund and discounting, USD Libor 3M, EUR discounting and EUR Euribor 3M. The curves are built in three units. The first one with the USD discounting curve calibrated to OIS, the second one with USD forward calibrated to IRS and the third one with both the EUR discounting and the EUR Euribor forward 3M calibrated to cross-currency swaps and EUR IRS. We use the hypothesis of independence of the market quotes of IRS Euribor 3M from the collateral currency to avoid convexity adjustment. This simplify the computations but not the dependency between the curves, which is the feature we want to emphasize in this section.

We look at the transition matrix for the EUR discounting. The transition matrix will depend potentially on all the other curves. Using the notation of our implementation, we have the summary dependency in Table 6.1. It has to be interpreted as follows: the generalized transition matrix depends on four

---

[1] The implementation we used is the OpenGamma Strata 1.0 library. It is open source and available on GitHub at https://github.com/OpenGamma/Strata

Table 6.1 Summarized representation of the dependency of the EUR with collateral in USD Fed Fund discounting curve to the other curves of the example

```
name=USD-DSCON-OIS, parameterCount=11,
name=EUR-DSC-XCCY, parameterCount=11,
name=EUR-EURIBOR3M-IRS, parameterCount=10,
name=USD-LIBOR3M-IRS, parameterCount=10
```

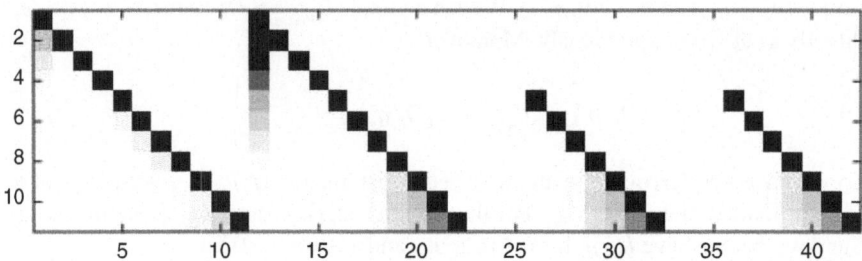

Fig. 6.1 Visual representation of the transition matrix of the EUR discounting for collateral in USD with Fed Fund rates

curves (with the listed names), for each curve the associated integer is the number of parameters of the curve. In this simplified example we have 42 parameters.

The total transition matrix for the last curve has 11 rows and 42 columns and is too large to be represented directly in this document. Instead we give a graphical representation of it in Fig. 6.1. The dark squares indicate a strong dependency (absolute value above 0.10) and the grey squares a low dependency down to the white squares where the dependency is 0.

The discounting curve in EUR for collateral in USD obviously depends on the cross-currency instruments (FX swaps and cross-currency swaps), but also on the USD discounting (OIS), USD single currency IRS and EUR single currency IRS.

Using this huge quantity of information and managing the dependencies is probably the main challenge in collateralized cross-currency trading and risk management. Algorithmic differentiation is helping not only to quicken the computation but also to clarify the dependencies.

Once the curve calibration process is finished, we suggest to store the inverse Jacobian matrix, at least the non-zero blocks of it, for later use. The information required for one curve, is the list of curves of which its parameters depends and the number of those parameters. In Strata, the object containing the curve name and its parameter count is called `CurveParameterSize`. The matrix itself is stored in a `JacobianCalibrationMatrix`. The simplified versions of the code are provided in Listing 6.1. An example of content for such an object is proposed in Table 6.1.

Listing 6.1   Curve Jacobian calibration matrix with reference to other curves and their market quote sizes

```
public class CurveParameterSize {
  /** The curve name. */
  private CurveName name;
  /** The number of parameters. */
  private int parameterCount;
  ...
}

public class JacobianCalibrationMatrix{
  /** The curve order. */
  private List<CurveParameterSize> order;
  /** The inverse Jacobian matrix. */
  private DoubleMatrix jacobianMatrix;
  ...
}
```

6.2   Model Calibration: Exact Calibration

We now discuss model calibration beyond the interest rate curves calibration. The content of this section is inspired from Henrard (2014a). Similar developments in the framework of PDE solutions in finance are proposed in Capriotti et al. (2015).

The method is first presented using a simple example, allowing simplified notation. The notation will be generalized later.

Suppose we have implemented the code for the function $f : \mathbb{R}^{p_a} \to \mathbb{R}^{p_z}$

$$z = f(a).$$

We suppose also that within the algorithm to compute f, there is an equation to solve. The algorithm is decomposed into

$$b = g_1(a)$$
$$c \text{ s. t. } g_2(b, c) = 0$$
$$z = g_3(c)$$

with $g_1 : \mathbb{R}^{p_a} \to \mathbb{R}^{p_b}$, $g_2 : \mathbb{R}^{p_b} \times \mathbb{R}^{p_c} \to \mathbb{R}^{p_c}$ and $g_3 : \mathbb{R}^{p_c} \to \mathbb{R}^{p_z}$. The second part of the algorithm is a multi-dimensional equation that must be solved. It is assumed that all intermediary functions g_i are differentiable. How to write the Algorithmic Differentiation version of this function? The middle part of the algorithm does not match the Single Assignment Code property described in Section 2.3.

Suppose that the adjoint versions of the functions g_i ($1 \leq i \leq 3$) are known but the adjoint version for the solver is unknown, this is the derivatives of the function that computes c from b are unknown. The implicit function theorem ensures – under certain mild conditions which are described in Appendix A.3 – that the process that produces c as function of b is actually differentiable and links its derivative to the derivatives of g_2. We denote by g_4 the implicit (and unknown) function associating c to b, this is the second part of the algorithm can be replaced by $g_4(b) = c$. We don't have code for the function g_4, we only know that the abstract version of the function exists around the point of interest. The implicit function theorem guarantees that g_4 exists and provides a way to compute its derivative based on the derivatives of the known function g_2 as

$$Dg_4(b) = -(D_2 g_2(b, c))^{-1} D_1 g_2(b, c).$$

Solving the equation $g_2(b, c) = 0$ is usually much more time-consuming than simply computing one value of g_2. The standard approach is to use a Newton-like algorithm and to repeat the valuation of g_2 with guessed values. In the implicit function theorem approach, using the adjoint version, there is no need to solve the equation again for the derivatives and there is no requirement to have adjoint version of the solver and its repeated guesses, only the derivatives of the function g_2 are used. Those derivatives are usually computed themselves by algorithmic differentiation. The adjoint method used in this way will give better results than the normal approach as there is no need to solve the equation for g_2 again. If the root-finding is a significant part of the computation time for f, the time required to compute the function f and all of its derivatives will be usually less than twice the time taken to calculate one value. The slower the root-finding problem is, the better on a relative basis for the enhanced AD approach.

The standard notation in AAD is to denote the derivative of the final value z with respect to an intermediate value x by $\bar{x}$. The literature uses different notations with regards to the transposition of $\bar{x}$; here we use

$$\bar{x} = (D_x z(x))^T,$$

this is the *bar* variables are column vectors if z is of dimension one, or matrices with p_x rows and p_z columns if z is of dimension greater than one.

The adjoint version of the algorithm is

$$\bar{z} = I \quad \text{(with } I \text{ the } p_z \times p_z \text{ identity)}$$
$$\bar{c} = (D_c g_3(c))^T \bar{z}$$
$$\bar{b} = (D_b g_4(b))^T \bar{c} = -\left((D_c g_2(b, c))^{-1} D_b g_2(b, c)\right)^T \bar{c}$$
$$\bar{a} = (D_a g_1(a))^T \bar{b}.$$

All the parts of the algorithm are now explicit. The implicit root-finding part has been replaced by an explicit matrix manipulation involving one matrix inversion.

6.2.1   Calibration Technique Description

One frequent part of exotic instruments pricing in finance is a *complex model calibration*. The process involves two model, one base model to price vanilla instruments and one complex model to price the exotic instrument. The process can be synthetizes as follows:

- The price of an *exotic instrument* is related to a specific basket of *vanilla instruments*;
- The price of these vanilla instruments is computed in a given *base model*;
- The *complex model parameters* are calibrated to fit the vanilla option prices from the base model. This step is usually done through a generic numerical equation solver; and
- The exotic instrument is then priced with the calibrated complex model.

   We want to differentiate the exotic price with respect to the parameters of the base model. In the bump and recompute approach, this corresponds to computing the risks with model recalibration. As the calibration can represent a major fraction of the total computation time of the procedure listed above, an alternative method is highly desirable.

   In this section, the generic AD and implicit function theorem method described in the first part of the section is applied to an interest rate model calibration case. Let $\mathrm{PV}_{\mathrm{Base}}^{\mathrm{Vanilla}}$ be the present values of a set of vanilla financial instruments used for calibration of the base model. The data required for the pricing are the yield curves in a multi-curve framework, which are denoted C, and market volatility parameters – for example SABR parameters or Black volatilities – for the base model. The parameters for the base model are denoted Θ. The exotic model provides a present value for the same vanilla options with the same curves but using a different set of parameters with different meaning. The set of parameters for the exotic model are denoted Φ. The pricing function for the vanilla options in the calibrated model is denoted $\mathrm{PV}_{\mathrm{Calibrated}}^{\mathrm{Vanilla}}$. The calibration procedure consists in finding the parameters Φ which solve

$$f(C, \Theta, \Phi) = 0. \tag{6.5}$$

For perfect calibration, the function is simply

$$f(C, \Theta, \Phi) = \mathrm{PV}_{\mathrm{Base}}^{\mathrm{Vanilla}}(C, \Theta) - \mathrm{PV}_{\mathrm{Calibrated}}^{\mathrm{Vanilla}}(C, \Phi).$$

Equation 6.5 will be multi-dimensional when there are several calibrating instruments. We suppose that there are as many calibration instruments as parameters to be calibrated in Φ.

   In practice, some models may have more free parameters than calibrating instruments. In this case the model parameters are constrained in such a way that there are the same number of degrees of freedom as the number of calibrating instruments. The second example below calibrates a two-factor LMM with many parameters. The parameters Φ used here are those degrees of freedom, not the original model parameters.

With the calibration procedure, we obtain implicit calibrated model parameters from the original model parameters and the curves:

$$\Phi = \Phi(C, \Theta).$$

The function is obtained through the equation solving procedure; there is no explicit solution or even explicit code that produces the function Φ directly from the inputs C and Θ.

The exotic option is priced from the calibrated model through the pricing $\text{PV}_{\text{Calibrated}}^{\text{Exotic}}(C, \Phi)$. With the implicit function above we can define

$$\text{PV}_{\text{Base}}^{\text{Exotic}}(C, \Theta) = \text{PV}_{\text{Calibrated}}^{\text{Exotic}}(C, \Phi(C, \Theta))$$

We are interested in the derivative of the exotic option with respect to the curves and the base model parameters Θ. The goal is to see all the risk, for the vanilla instruments and the exotic instruments written in term of the market quotes used in the liquid vanilla market. This gives a coherent view of the risk for a portfolio using heterogeneous models.

With the derivative versions of $\text{PV}_{\text{Calibrated}}^{\text{Exotic}}$, we can compute the derivatives

$$D_C \text{PV}_{\text{Calibrated}}^{\text{Exotic}} \text{ and } D_\Phi \text{PV}_{\text{Calibrated}}^{\text{Exotic}}.$$

The quantities of interest are

$$D_C \text{PV}_{\text{Base}}^{\text{Exotic}} \text{ and } D_\Theta \text{PV}_{\text{Base}}^{\text{Exotic}}.$$

Through composition we have

$$D_C \text{PV}_{\text{Base}}^{\text{Exotic}}(C, \Theta) = D_C \text{PV}_{\text{Calibrated}}^{\text{Exotic}}(C, \Phi(C, \Theta))$$
$$+ D_\Phi \text{PV}_{\text{Calibrated}}^{\text{Exotic}}(C, \Phi(C, \Theta)) D_C \Phi(C, \Theta),$$

and

$$D_\Theta \text{PV}_{\text{Base}}^{\text{Exotic}}(C, \Theta) = D_\Phi \text{PV}_{\text{Calibrated}}^{\text{Exotic}}(C, \Phi(C, \Theta)) D_\Theta \Phi(C, \Theta).$$

The quantities $D_C \Phi$ and $D_\Theta \Phi$ are still unknown at this stage. Using the implicit function theorem, the function Φ is differentiable and its derivatives can be computed from the derivative of f:

$$D_\Theta \Phi(C, \Theta) = -(D_\Phi f(C, \Theta, \Phi(C, \Theta)))^{-1} D_\Theta f(C, \Theta, \Phi(C, \Theta))$$

and

$$D_C\Phi(C, \Theta) = -(D_\Phi f(C, \Theta, \Phi(C, \Theta)))^{-1} D_C f(C, \Theta, \Phi(C, \Theta)).$$

In the perfect calibration case, those equations reduce to

$$D_\Theta\Phi(C, \Theta) = \left(D_\Phi PV_{\text{Calibrated}}^{\text{Vanilla}}(C, \Phi(C, \Theta))\right)^{-1} D_\Theta PV_{\text{Base}}^{\text{Vanilla}}(C, \Theta)$$

and

$$D_C\Phi(C, \Theta) = \left(D_\Phi PV_{\text{Calibrated}}^{\text{Vanilla}}(C, \Phi(C, \Theta))\right)^{-1}$$
$$\left(D_C PV_{\text{Base}}^{\text{Vanilla}}(C, \Theta) - D_C PV_{\text{Calibrated}}^{\text{Vanilla}}(C, \Phi(C, \Theta))\right).$$

We have described the theoretical results leading to an efficient implementation of sensitivity computation through the calibration procedure. The procedure only requires the implementation of the derivatives of the PV functions and a little bit of linear algebra.

6.2.2   Examples

In line with the above technique, we would like to price and compute the sensitivities of exotic swaptions in a physical delivery SABR framework. For all the required pricing algorithms the adjoint versions have been implemented[2].

6.2.2.1   Cash Swaptions in the Hull-White Model

In the first example our *exotic* instrument is a cash-settled swaption and our vanilla basket is composed of a unique physical delivery swaption. The base model is a SABR model on the swap rate. The curve framework is multi-curve (one discounting curve and one forward curve). The model parameters Θ are the SABR parameters α, ρ and ν (β is set to 0.50). The calibrated model is a Hull-White one factor (extended Vasicek) model with constant volatility. The parameter of the Hull-White model to calibrate is the constant volatility. The pricing algorithm in the Hull-White model for the physical delivery swaption is described in Henrard (2003), and the pricing algorithm used for the cash-settled swaption is the efficient approximation described in Henrard (2010a).

For this example, we use a 1Y × 9Y swaption on an annual vs 6m Euribor swap. There are three SABR sensitivities (α, ρ, and ν) and 38 rate sensitivities (19 on each curve). The performance results are provided in Table 6.2. The computation of the

---

[2] The implementations used for the performance figures are those in the OpenGamma analytics library. The computations are done on a Mac Pro 3.2 GHz Quad-core.

Table 6.2 Performance for different approaches to derivatives computations: cash settled swaption in Hull-White one factor model

Risk type	Approach	Price time	Risks time	Total
SABR	Finite difference	1.00	3×1.00	4.00
SABR	AAD and implicit function	1.00	0.28	1.28
Curve	Finite difference	1.00	38×1.00	39.00
Curve	AAD and implicit function	1.00	0.56	1.56
Curve and SABR	Finite difference	1.00	41×1.00	42.00
Curve and SABR	AAD and implicit function	1.00	0.83	1.83

Times relative to the pricing time. The pricing time is 0.45 second for 1000 swaptions.

three SABR derivatives add less than 30% to the pricing computation time in this approach; a one-sided finite difference computation would have add at least 300%.

The same comparison was performed for the interest rate sensitivities. The finite difference would require at least 39 price time (3900%). The proposed approach adds only 0.55 price time (55%). In total, the proposed algorithm is around 23 times faster than a finite difference approach and is numerically more stable.

To partly compare the above numbers with the results of Schlenkirch (2012), we also report figures for a Hull-White one factor model with piecewise constant volatility. The set-up in the above paper is different but the underlying model is similar. The computation time for the finite difference and adjoint evaluations of the Jacobian with respect to the piecewise constant volatility for a 30Y and 100Y swap (annual volatility dates) is provided. The Jacobian is the derivative of all European swaption prices with respect to all volatilities in the model. The 30Y Jacobian computation requires 0.110 second by finite difference and less than 0.015 second by algorithmic differentiation. This is approximately 14% of the runtime and is in line with Schlenkirch (2012) figures (20%). The corresponding figures for the 100Y case are 20.2 s and 0.42 s (2%).

6.2.2.2   Amortized Swaptions in LMM

In this example, the exotic instrument is an amortized European swaption (i.e. a swaption with decreasing notional), and the vanilla basket is composed of vanilla European swaptions with same expiry and increasing maturities. The amortized swaption has a 10Y maturity and yearly amortization. The calibrating instruments are ten vanilla swaptions with yearly maturities between 1Y and 10Y and same strike as the amortized swaption.

The base model is a SABR model on each vanilla swaption. The complex model is a two-factor LMM with displaced diffusion and Libor period of six months. The pricing method for the vanilla and the amortized swaption is the efficient approximation described in Henrard (2010b).

The calibration is performed as follows: for each yearly period the weights of the different parameters (four in each year) are fixed. The calibration is done by

Table 6.3 Performance for different approaches to derivatives computations: amortized swaption in the LMM

Risk type	Approach	Price time	Risks time	Total
SABR	Finite difference	1.00	30×1.00	31.00
SABR	AAD and implicit function	1.00	0.18	1.18
Curve	Finite difference	1.00	42×1.00	43.00
Curve	AAD and implicit function	1.00	0.74	1.74
Curve and SABR	Finite difference	1.00	72×1.00	73.00
Curve and SABR	AAD and implicit function	1.00	0.75	1.75

Times relative to the pricing time. The valuation time is 0.425 second for 250 swaptions.

multiplying those weights by a common factor. The parameter Φ in the previous section are the multiplicative factors (10 in total), even if in practice the derivatives with all the model parameters (40 in total) are computed as an intermediary step.

The computation time results for the SABR and curve sensitivities are reported in Table 6.3. There are 30 SABR sensitivities (α, ρ, ν for 10 vanilla swaptions). In the described approach, the 30 sensitivities add only 20% to the computation time with respect to the calibration and price computation .

There are 42 curve sensitivities (two curves, semi-annual payments over 10 years). The computation of the 42 sensitivities takes only 74% of the price time. In total, the AAD approach is approximately 2.5% of the time required by finite difference. Note that computing the curve and SABR sensitivities take approximately the same amount of time as computing the curve sensitivities alone as most of the computations are common.

Similar results for amortized swaptions of different maturities and with different numbers of calibrating instruments are reported in Fig. 6.2. The ratios between the present value and sensitivities time and the present value time are reported for the finite difference approach and Adjoint Algorithmic Differentiation approach using the implicit function method described in the previous section. The implicit function AAD method ratio is almost independent of the number of sensitivities. In all cases but the 30Y swaptions (where 212 sensitivities are calculated), the ratio is below two. For the finite differences, the ratio is above 200, a gain of more than a factor 100. This mean the computation time is reduced from one hour to less than one minute.

6.3   Model Calibration: Least-Square

In this section we develop techniques similar to the one developed in Section 6.2 but for the case where the calibration is not a root-finding calibration but a least square calibration.

Here we consider the case were the parameters Φ are obtained through a (weighted) least square process. Suppose that there are n calibrated parameters in Φ and $m \geq n$ instruments for the calibration process. The weights associated to

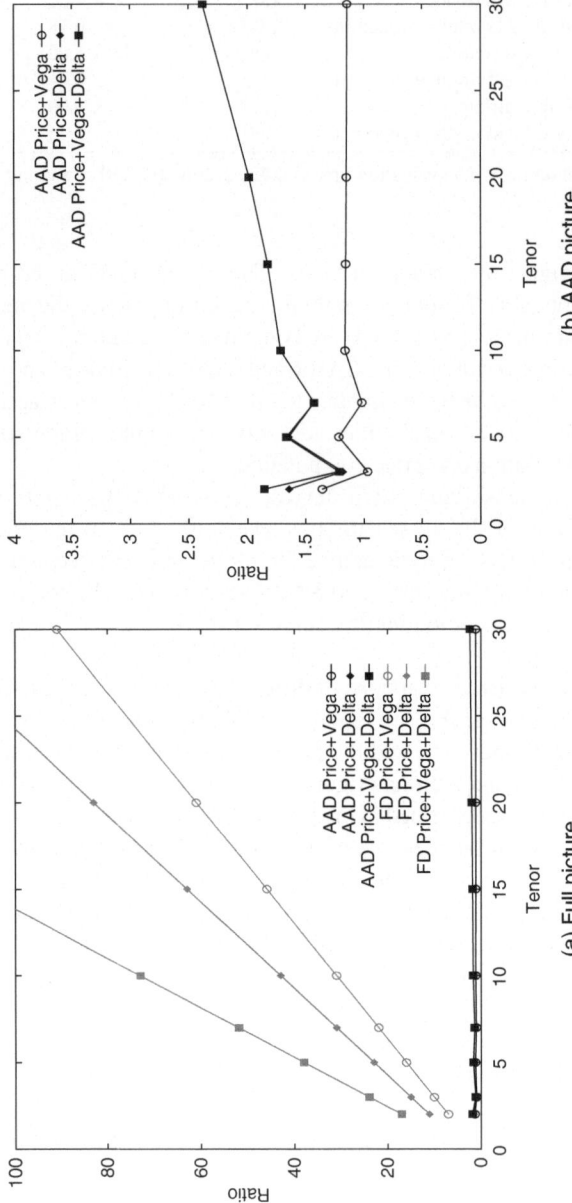

Fig. 6.2 Computation time ratios (present value and sensitivities time to present value time) for the finite difference and AAD methods. The *vega* represents the derivatives with respect to the SABR parameters; the *delta* represents the derivatives with respect to the curves. The AAD method uses the implicit function approach. Figures for annually amortized swaptions in a LMM calibrated to vanilla swaptions in valued with the SABR model

each instrument are fixed and denoted $(w_i)_{i=1,...,m}$. The calibration parameters are defined as

$$\Phi_0 = \arg\min_{\Phi \in \mathbb{R}^n} h(C_0, \Theta_0, \Phi)$$
$$= \arg\min_{\Phi} \sum_{i=1,...,m} w_i \left(PV_{\text{Base}}^{\text{Vanilla}}(i, C_0, \Theta_0) - PV_{\text{Calibrated}}^{\text{Vanilla}}(i, C_0, \Phi) \right)^2. \quad (6.6)$$

At the minimum Φ_0, the derivatives of h with respect to Φ are 0:

$$f(C_0, \Theta_0, \Phi_0) = D_\Phi h(C_0, \Theta_0, \Phi_0) = 0.$$

The minimum satisfies Eq. 6.5 with the function f defined above. This last equation is a n unknown and n equation system. The above derivatives can be computed explicitly as

$$D_\Phi h(C, \Theta, \Phi)$$
$$= -2 \sum_{i=1,...,m} w_i \left(PV_{\text{Base}}^{\text{Vanilla}}(i, C, \Theta) - PV_{\text{Calibrated}}^{\text{Vanilla}}(i, C, \Phi) \right) D_\Phi PV_{\text{Calibrated}}^{\text{Vanilla}}(i, C, \Phi).$$

With the calibration procedure, we obtain calibrated model parameters from the original model parameters and the curves:

$$\Phi_0 = \Phi(C_0, \Theta_0).$$

The parameters are obtained through the optimization procedure; there is no explicit solution or even explicit code that produces those parameters directly. We suppose that f is regular enough and that $D_\Phi f(C_0, \Theta_0, \Phi_0)$ is invertible, so we can apply the implicit function theorem.

The implicit function theorem states that there exists a function $\Phi(C, \Theta)$ such that

$$f(C, \Theta, \Phi(C, \Theta)) = 0$$

for (C, Θ) close to (C_0, Θ_0) and there are no other solution in a neighborhood. We still need to prove that the function $\Phi(C, \Theta)$ gives a minimum of the original problem (6.6) and not only a point with 0 derivatives, like a saddle point.

Let m_0 denote the minimum value of (6.6) at Φ_0, this is $m_0 = h(C_0, \Theta_0, \Phi_0)$. As Φ_0 is a minimum, $D_\Phi f(C_0, \Theta_0, \Phi_0)$ is positively defined. As it is invertible, it is strictly positively defined. Using those properties, one can show that there exists a $\epsilon > 0$ and a sphere around Φ_0 such that $h(C_0, \Theta_0, \Phi) > m_0 + 3\epsilon$ for Φ on the sphere. As h is continuous, for (C, Θ) close enough to (C_0, Θ_0) and Φ on the sphere, $h(C, \Theta, \Phi) > m_0 + 2\epsilon$. On the other side, $h(C, \Theta, \Phi(C, \Theta)) < m_0 + \epsilon$ for (C, Θ) close enough to (C_0, Θ_0). This proves that h has a minimum in the interior of the disk in the Φ dimension. Being in the interior, the minimum has zero derivatives. From the result of the implicit function theorem, $\Phi(C, \Theta)$ is the only zero. This proves that the implicit function $\Phi(C, \Theta)$ is not only a zero of the derivative function f but also a minimum of the least square problem.

The exotic option is priced from the calibrated model through the pricing $\mathrm{PV}_{\mathrm{Calibrated}}^{\mathrm{Exotic}}(C, \Phi)$. With the implicit function above we can define

$$\mathrm{PV}_{\mathrm{Base}}^{\mathrm{Exotic}}(C, \Theta) = \mathrm{PV}_{\mathrm{Calibrated}}^{\mathrm{Exotic}}(C, \Phi(C, \Theta))$$

We are interested in the derivative of the exotic option with respect to the curves and the base model parameters Θ.

The quantities of interest are

$$D_C \mathrm{PV}_{\mathrm{Base}}^{\mathrm{Exotic}} \text{ and } D_\Theta \mathrm{PV}_{\mathrm{Base}}^{\mathrm{Exotic}}.$$

With the AD versions of $\mathrm{PV}_{\mathrm{Calibrated}}^{\mathrm{Exotic}}$, we can compute the derivatives

$$D_C \mathrm{PV}_{\mathrm{Calibrated}}^{\mathrm{Exotic}} \text{ and } D_\Phi \mathrm{PV}_{\mathrm{Calibrated}}^{\mathrm{Exotic}}.$$

Through composition we have

$$D_C \mathrm{PV}_{\mathrm{Base}}^{\mathrm{Exotic}}(C_0, \Theta_0)$$
$$= D_C \mathrm{PV}_{\mathrm{Calibrated}}^{\mathrm{Exotic}}(C_0, \Phi(C_0, \Theta_0)) + D_\Phi \mathrm{PV}_{\mathrm{Calibrated}}^{\mathrm{Exotic}}(C_0, \Phi(C_0, \Theta_0)) D_C \Phi(C_0, \Theta_0),$$

and

$$D_\Theta \mathrm{PV}_{\mathrm{Base}}^{\mathrm{Exotic}}(C_0, \Theta_0) = D_\Phi \mathrm{PV}_{\mathrm{Calibrated}}^{\mathrm{Exotic}}(C_0, \Phi(C_0, \Theta_0)) D_\Theta \Phi(C_0, \Theta_0).$$

Where $D_C \Phi$ and $D_\Theta \Phi$ are yet unknown. Using the implicit function theorem, the function Φ is differentiable and its derivatives can be computed from the derivative of f:

$$D_\Theta \Phi(C_0, \Theta_0) = -(D_\Phi f(C_0, \Theta_0, \Phi(C_0, \Theta_0)))^{-1} D_\Theta f(C_0, \Theta_0, \Phi(C_0, \Theta_0))$$

and

$$D_C \Phi(C_0, \Theta_0) = -(D_\Phi f(C_0, \Theta_0, \Phi(C_0, \Theta_0)))^{-1} D_C f(C_0, \Theta_0, \Phi(C_0, \Theta_0)).$$

We need to describe $D_X f$.

$$D_\Theta f(C, \Theta, \Phi) = D_\Theta D_\Phi h(C, \Theta, \Phi)$$
$$= -2 \sum_{i=1,\ldots,n} w_i D_\Phi^T \mathrm{PV}_{\mathrm{Calibrated}}^{\mathrm{Vanilla}}(i, C, \Phi) D_\Theta \mathrm{PV}_{\mathrm{Base}}^{\mathrm{Vanilla}}(i, C, \Theta).$$

$$D_C f(C, \Theta, \Phi) = -2 \sum_{i=1,\dots,n} w_i D_\Phi^T \mathrm{PV}_{\mathrm{Calibrated}}^{\mathrm{Vanilla}}(i, C, \Phi) \left(D_C \mathrm{PV}_{\mathrm{Base}}^{\mathrm{Vanilla}}(i, C, \Theta) \right.$$

$$\left. -D_C \mathrm{PV}_{\mathrm{Calibrated}}^{\mathrm{Vanilla}}(i, C, \Phi) \right) - 2 \sum_{i=1,\dots,n} w_i \left(\mathrm{PV}_{\mathrm{Base}}^{\mathrm{Vanilla}}(i, C, \Theta) \right.$$

$$\left. -\mathrm{PV}_{\mathrm{Calibrated}}^{\mathrm{Vanilla}}(i, C, \Phi) \right) D_C D_\Phi \mathrm{PV}_{\mathrm{Calibrated}}^{\mathrm{Vanilla}}(i, C, \Phi)$$

$$D_\Phi f(C, \Theta, \Phi)$$

$$= D_\Phi D_\Phi h(C, \Theta, \Phi) = 2 \sum_{i=1,\dots,n} w_i D_\Phi \mathrm{PV}_{\mathrm{Calibrated}}^{\mathrm{Vanilla}}(i, C, \Phi) D_\Phi^T \mathrm{PV}_{\mathrm{Calibrated}}^{\mathrm{Vanilla}}(i, C, \Phi)$$

$$-2 \sum_{i=1,\dots,n} w_i \left(\mathrm{PV}_{\mathrm{Base}}^{\mathrm{Vanilla}}(i, C, \Theta) - \mathrm{PV}_{\mathrm{Calibrated}}^{\mathrm{Vanilla}}(i, C, \Phi) \right) D_\Phi^2 \mathrm{PV}_{\mathrm{Calibrated}}^{\mathrm{Vanilla}}(i, C, \Phi).$$

The *annoying* parts are the second order derivatives parts. Usually the first order derivatives are implemented in AD frameworks but not the second order one. Fortunately in the above formula, all the second order derivatives are multiplied by $\mathrm{PV}_{\mathrm{Base}}^{\mathrm{Vanilla}}(i, C, \Theta) - \mathrm{PV}_{\mathrm{Calibrated}}^{\mathrm{Vanilla}}(i, C, \Phi)$ which is small when the calibrated model can match the base prices well enough. Based on that, we can use the following approximations:

$$D_C f(C, \Theta, \Phi)$$

$$\simeq -2 \sum_{i=1,\dots,n} w_i D_\Phi^T \mathrm{PV}_{\mathrm{Calibrated}}^{\mathrm{Vanilla}}(i, C, \Phi) \left(D_C \mathrm{PV}_{\mathrm{Base}}^{\mathrm{Vanilla}}(i, C, \Theta) \right.$$

$$\left. -D_C \mathrm{PV}_{\mathrm{Calibrated}}^{\mathrm{Vanilla}}(i, C, \Phi) \right)$$

and

$$D_\Phi f(C, \Theta, \Phi) \simeq 2 \sum_{i=1,\dots,n} w_i D_\Phi^T \mathrm{PV}_{\mathrm{Calibrated}}^{\mathrm{Vanilla}}(i, C, \Phi) D_\Phi \mathrm{PV}_{\mathrm{Calibrated}}^{\mathrm{Vanilla}}(i, C, \Phi).$$

This is very similar to the approximation done in computing Hessian as described in (Press et al., 1988, Section 14.4).

6.3.1   Least Square Examples

In this section we analyze an example similar to the second one of the previous example section.

The calibrated model is a Libor Market Model with displaced diffusion. We calibrate two parameters for each maturity: the volatility and the displacement parameters. The volatility parameter guides the general level of the smile while the

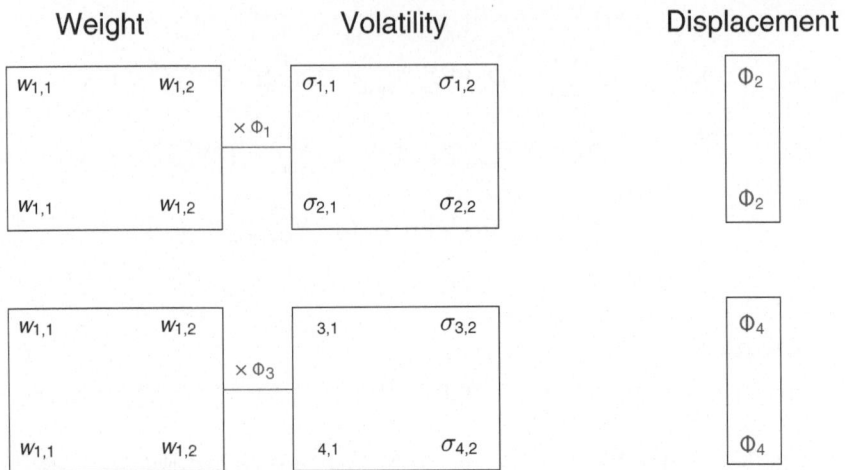

Fig. 6.3 Representation of LMM calibration for a 2 years amortised swaption. Each yearly volatility block is multiplied by a common multiplicative factor and each yearly displacement block contains the same number

displacement parameter commands the skew of the smile. We calibrate the two parameters by a least square approach on the prices of swaptions with several strikes. In the tests we use between 2 and 6 strikes.

The calibration is done for each yearly block on a multiplicative factor to given weights to obtain volatilities and on a shared displacement. The graphical representation of the calibration for a 2 year problem is provided in Fig. 6.3.

The more difficult the calibration is, the better the results of AD with implicit function will be on a relative basis. The algorithmic differentiation with implicit theorem method is using the already computed calibration in the sensitivity.

6.3.1.1   Reference

The reference example has a tenor of 5 years and there are five annual calibrations. The calibration on each tenor is done on three strikes ($-100, 0, +100$) bps from ATM. In the finite difference approach, the ratio is roughly 3 (SABR) + 4 (semi-annual payments with 2 curve) for each years. For a 5 years tenor, the finite difference ratio is around 36. The ratio obtained in practice in this example through AD with implicit function is 1.40.

6.3.1.2   Tenors

We run the same test with several tenors, between 2 years and 30 years. The calibration is similarly annual on three swaptions for each calibration date. In a finite difference the ratios would increase roughly linearly with the tenor. Figure 6.4 reports the results.

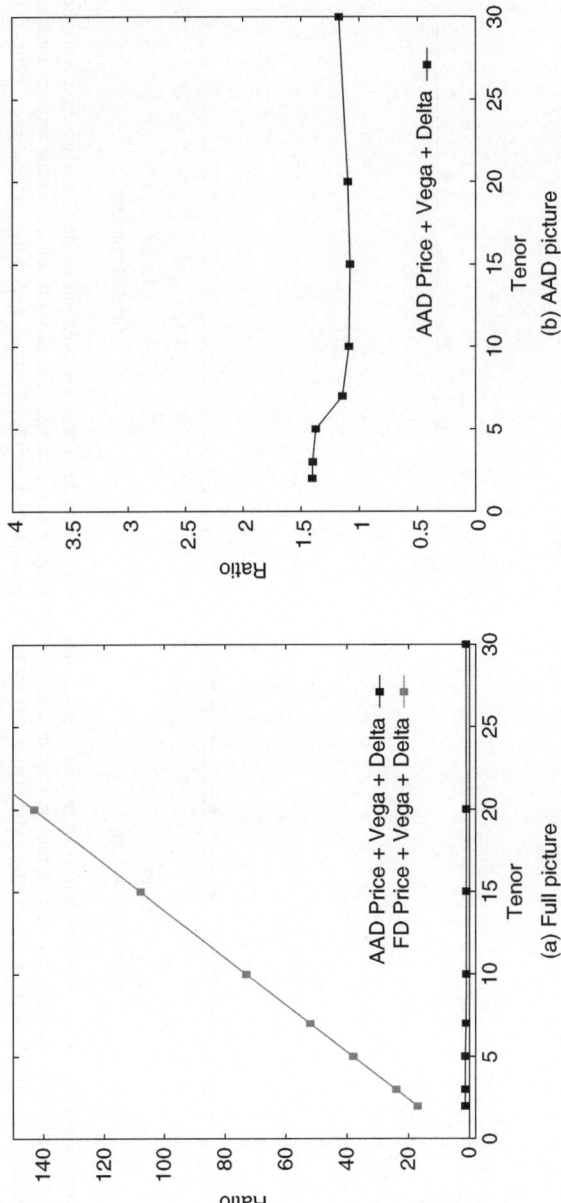

Fig. 6.4 Computation time ratios (price and sensitivities time to price time) for the finite difference and AAD methods. The *vega* represents the derivatives with respect to the SABR parameters; the *delta* represents the derivatives with respect to the curves. The AAD method uses the implicit function approach. Figures for annually amortised swaptions in a LMM calibrated yearly to three vanilla swaptions in SABR

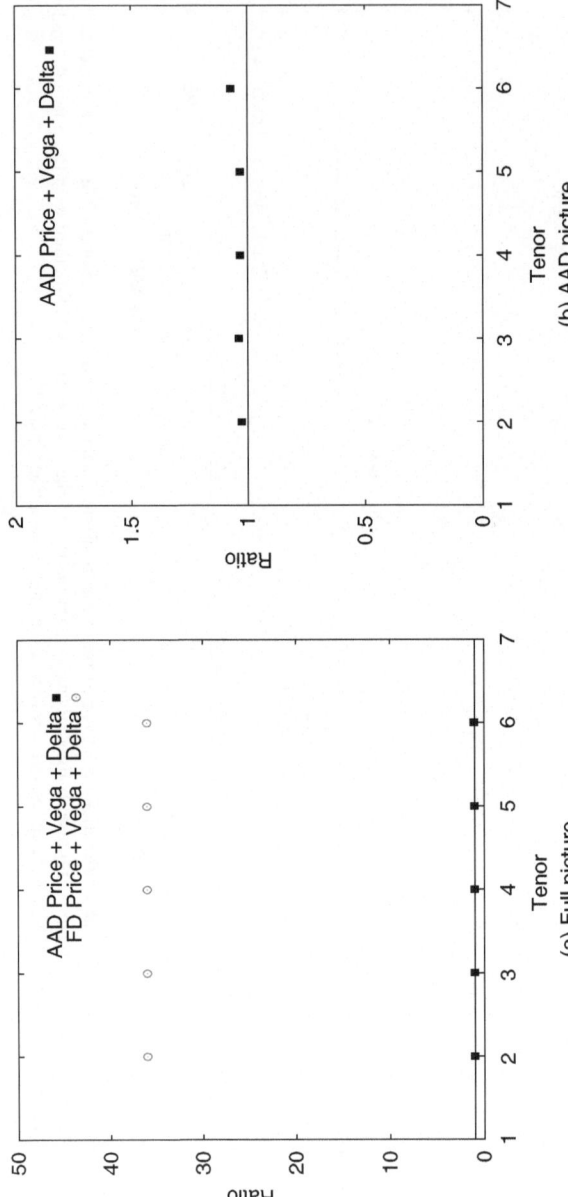

Fig. 6.5 Computation time ratios (price and sensitivities time to price time) for the finite difference and AAD methods. The *vega* represents the derivatives with respect to the SABR parameters; the *delta* represents the derivatives with respect to the curves. The AAD method uses the implicit function approach. Figures for annually amortised swaptions in a LMM calibrated yearly to the given number of vanilla swaptions with different strikes in SABR

As in the previous examples, the metric to analyze the efficiency is the ratio between price and derivatives time and price time. The derivatives are composed of the SABR and curve sensitivities.

The linear increase of the ratios with the tenors is obvious for the finite difference method. The AD with implicit function method achieves a relatively constant ratio which is barely above 1 and well below 1.5. This is well below the theoretical upper bound of $\omega_A \in [3, 4]$. In the case of the 20-year swaption, the gain between the finite difference and optimized AD with implicit function is around 100. In practice this is reducing the computation time from 1 hour 40 minutes to 1 minute.

6.3.1.3 Strikes

In this part of the analysis, we go back to a 5 year amortised swaption with annual calibration. We run similar tests with calibrations sets for each periods with 2–6 strikes. Figure 6.5 reports the results.

As can be seen, the ratios are independent of the number of calibrating strikes for the finite difference and the AD versions. The ratio obtained in the AD with implicit function case is around 1.1, which is well below the theoretical upper bound of $\omega_A \in [3, 4]$. The ratio for the finite difference approach is around 36 and independent of the number of strikes.

Bibliography

Capriotti, L., Jiang, Y., and Macrina, A. (2015). Real-time risk management: An AAD-PDE approach. *Journal of Financial Engineering*, 2(4):1–31.

Henrard, M. (2003). Explicit bond option and swaption formula in Heath-Jarrow-Morton one-factor model. *International Journal of Theoretical and Applied Finance*, 6(1):57–72.

Henrard, M. (2010a). Cash-Settled Swaptions: How Wrong are We? Working paper series 1703846, SSRN. Available at SSRN: http://ssrn.com/abstract=1703846.

Henrard, M. (2010b). Swaptions in Libor Market Model with local volatility. *Wilmott Journal*, 2(3):135–154.

Henrard, M. (2014a). Adjoint algorithmic differentiation: Calibration and implicit function theorem. *Journal of Computational Finance*, 17(4):37–47.

Henrard, M. (2014b). *Interest rate modelling in the multi-curve framework: Foundations, evolution and implementation*. Applied Quantitative Finance. London:Palgrave Macmillan. ISBN: 978-1-137-37465-3.

Press, W. H., Flannery, B. P., Teukolsky, S. A., and Vetterling, W. T. (1988). *Numerical Recipes in C: The Art of Scientific Computing*. Cambridge:Cambridge University Press.

Schlenkirch, S. (2012). Efficient calibration of the Hull-White model. *Optimal Control Applications and Methods*, 33(3):352–362.

Appendix A
Mathematical Results

The most powerful force in the universe is composition. – What we need is implicit.

A.1 Derivative Notations

Definition A.1 (Derivative) A function $f : \mathbb{R}^m \to \mathbb{R}^n; x \mapsto f(x)$ is said to be *differentiable* at a point $x_0 \in \mathbb{R}^m$ if f is defined on that point and there exist a linear function $Df(x_0) : \mathbb{R}^m \to \mathbb{R}^n$ such that

$$\lim_{\epsilon \to 0; \epsilon \in \mathbb{R}^m} \frac{f(x_0 + \epsilon) - (f(x_0) + Df(x_0)(\epsilon))}{|\epsilon|} = 0.$$

The linear function $Df(x_0)$ is called the *derivative* of f in x_0.

If the input vector is split in several sub-vectors, like in $(a_1, a_2) \mapsto z = f(a_1, a_2)$, the partial derivative with respect of one of the partial vector is denoted $D_i f(a_1, a_2)$ or $D_{a_i} f(a_1, a_2)$ with $i = 1$ or 2.

The derivative is a linear function with $Df(x) \in \mathcal{L}(\mathbb{R}^m, \mathbb{R}^n)$. The derivative is represented by a $n \times m$ matrix (n rows, m columns). For column vectors a and ϵ, we will often used the approximation

$$f(a + \epsilon) \sim f(a) + Df(a) \cdot \epsilon.$$

A.2 Derivative of Composition

Theorem A.1 Let $g_1 : \mathbb{R}^{p_a} \to \mathbb{R}^{p_b}$ be differentiable in a and $g_2 : \mathbb{R}^{p_b} \to \mathbb{R}^{p_z}$ be differentiable in $g_1(a)$. Then the function

$$(g_2 \circ g_1) : \mathbb{R}^{p_a} \to \mathbb{R}^{p_z}; a \mapsto g_2(g_1(a))$$

is differentiable in a and

$$D(g_2 \circ g_1)(a) = D(g_2(g_1(a)) \cdot Dg_1(a).$$

© The Author(s) 2017
M. Henrard, *Algorithmic Differentiation in Finance Explained*,
Financial Engineering Explained, DOI 10.1007/978-3-319-53979-9

The derivative operator transform the composition of function into a composition of linear function, represented by a matrix multiplication.

This result has a positive impact on the efficiency of Algorithmic Differentiation. The complex process of composition is replace at the derivative level by a matrix multiplication which can be performed very efficiently by computers.

A.3   Implicit Function Theorem

Theorem A.2   Le $f : \mathbb{R}^{n+m} \to \mathbb{R}^m$ be continuously differentiable. If (x_0, y_0) is such that

$$f(x_0, y_0) = 0$$

and if $D_y f(x_0, y_0)$ is invertible, then, in a neighborhood $X \times Y$ of (x_0, y_0), there is a (implicit) function g such that $f(x, g(x)) = 0$ for $x \in X$, $\{(x, g(x)) : x \in X\} = \{(x, y) \in X \times Y : f(x, y) = 0\}$, g is differentiable in x_0 and

$$Dg(x_0) = -(D_2 f(x_0, y_0))^{-1} D_1 f(x_0, y_0).$$

See for example Mawhin (1997) for more details and a proof of the result.

This result has also a positive impact on the efficiency of Algorithmic Differentiation. The complex process of root-finding is replaced at the derivative level by a matrix inversion which can also be performed very efficiently by computers.

Bibliography

Mawhin, J. (1997). *Analyse: Fondements, Techniques, Evolution*. Louvain-la-Neuve:De Boeck.

Index

© The Author(s) 2017
M. Henrard, *Algorithmic Differentiation in Finance Explained*,
Financial Engineering Explained, DOI 10.1007/978-3-319-53979-9

Index

Lecture Notes in Computer Science   14666

Founding Editors

Gerhard Goos
Juris Hartmanis

Editorial Board Members

The series Lecture Notes in Computer Science (LNCS), including its subseries Lecture Notes in Artificial Intelligence (LNAI) and Lecture Notes in Bioinformatics (LNBI), has established itself as a medium for the publication of new developments in computer science and information technology research, teaching, and education.

LNCS enjoys close cooperation with the computer science R & D community, the series counts many renowned academics among its volume editors and paper authors, and collaborates with prestigious societies. Its mission is to serve this international community by providing an invaluable service, mainly focused on the publication of conference and workshop proceedings and postproceedings. LNCS commenced publication in 1973.

Marek Michalewicz · John Gustafson ·
Himeshi De Silva
Editors

Next Generation Arithmetic

5th International Conference, CoNGA 2024
Sydney, NSW, Australia, February 20–21, 2024
Proceedings

 Springer

Editors
Marek Michalewicz ⓘ
National Supercomputing Centre Singapore
Singapore, Singapore

John Gustafson ⓘ
Arizona State University
Tempe, AZ, USA

Himeshi De Silva
Agency for Science, Technology
and Research
Singapore, Singapore

ISSN 0302-9743   ISSN 1611-3349 (electronic)
Lecture Notes in Computer Science
ISBN 978-3-031-72708-5   ISBN 978-3-031-72709-2 (eBook)
https://doi.org/10.1007/978-3-031-72709-2

This Springer imprint is published by the registered company Springer Nature Switzerland AG
The registered company address is: Gewerbestrasse 11, 6330 Cham, Switzerland

If disposing of this product, please recycle the paper.

Preface

Many of the earliest electronic computers used decimal arithmetic, in imitation of the way humans perform calculations. Eventually engineers realized the massive gains in efficiency of using binary arithmetic. Similarly, computers expressed real numbers using "floating-point arithmetic," in imitation of the scientific notation that humans use. The most common real-number format in use today is based on choices made by Intel engineer John Palmer for a coprocessor designed half a century ago, later codified as IEEE Std 754™.

The break from this legacy began seven years ago with the introduction of the *posit* format, a format designed based on mathematics instead of an imitation of human notation. The paper that introduced that format has had hundreds of citations and led to posit-based projects around the world, as well as the annual Conference on Next-Generation Arithmetic (CoNGA).

These are the Proceedings of the Fifth CoNGA meeting, which was held as a standalone virtual conference 20th February 2024. The invited keynote presentation was by executives from Calligo Technologies Inc. of Bangalore, India, who announced the world's first custom VLSI embodiment of posit arithmetic in a multicore processor, with a complete software stack, including applications ready-to-run. Their pioneering effort makes it effortless to compare the answer quality of next-generation arithmetic with the legacy floating-point format.

Among the several outstanding submissions that were accepted and appear in this volume, a standout work is the paper by Laslo Hunhold, "Beating Posits at Their Own Game: Takum Arithmetic." Hunhold's invention of a more compact way to express scale factors fits within the posit framework but appears superior to both IEEE floats and the original posit definition.

March 2024

John Gustafson
Marek Michalewicz

Organization

Co-chairs

John L. Gustafson   Arizona State University, USA
Marek Michalewicz   National Supercomputing Centre, Singapore
Himeshi De Silva   Agency for Science, Technology and Research
     (A*STAR), Singapore

Program Chair

Himeshi De Silva   Agency for Science, Technology and Research
     (A*STAR), Singapore

Program Committee

John Gustafson   Arizona State University, USA
Marek Michalewicz   National Supercomputing Centre, Singapore
Himeshi De Silva   A*STAR, Singapore
Marco Cococcioni   University of Pisa, Italy
Laslo Hunhold   University of Cologne, Germany
Peter Lindstrom   LLNL, USA
Roman Iakymchuk   Umeå University, Sweden
Glenn Matlin   Georgia Institute of Technology, USA
Shin Yee Chung   National Supercomputing Centre, Singapore
Akshat Ramachandran   Veermata Jijabai Technological Institute, India
Guillermo Botella   Universidad Complutense de Madrid, Spain
Vassil Dimitrov   Lemurian Labs, USA

Contents

Beating Posits at Their Own Game:
Takum Arithmetic

Laslo Hunhold[(✉)] [ID]

Parallel and Distributed Systems Group, University of Cologne, Cologne, Germany
hunhold@uni-koeln.de

Abstract. Recent evaluations have highlighted the tapered posit number format as a promising alternative to the uniform precision IEEE 754 floating-point numbers, which suffer from various deficiencies. Although the posit encoding scheme offers superior coding efficiency at values close to unity, its efficiency markedly diminishes with deviation from unity. This reduction in efficiency leads to suboptimal encodings and a consequent diminution in dynamic range, thereby rendering posits suboptimal for general-purpose computer arithmetic.

This paper introduces and formally proves 'takum' as a novel general-purpose logarithmic tapered-precision number format, synthesising the advantages of posits in low-bit applications with high encoding efficiency for numbers distant from unity. Takums exhibit an asymptotically constant dynamic range in terms of bit string length, which is delineated in the paper to be suitable for a general-purpose number format. It is demonstrated that takums either match or surpass existing alternatives. Moreover, takums address several issues previously identified in posits while unveiling novel and beneficial arithmetic properties.

Keywords: takum arithmetic · tapered number format · logarithmic number system · dynamic range · posit arithmetic

1 Introduction

The fundamental premise of machine number systems is to effectively represent numerical values within computers using bit strings. This paper assumes the reader's acquaintance with two prominent methodologies: the IEEE 754 floating-point format (refer to [12]) and posits (refer to [11]). For a comprehensive discussion on both, readers are directed to [8]. The assessment of merits and drawbacks within a number system is contingent upon the selected criteria. To lay the groundwork, we will outline a possible set of essential properties of an ideal number system in the subsequent discussion.

1.1 GUSTAFSON Criteria

GUSTAFSON outlines ten criteria for assessing a robust number system in [8]. We slightly extend their scope and designate specific keywords, yielding the subsequent list of properties:

© The Author(s), under exclusive license to Springer Nature Switzerland AG 2024
M. Michalewicz et al. (Eds.): CoNGA 2024, LNCS 14666, pp. 1–51, 2024.
https://doi.org/10.1007/978-3-031-72709-2_1

1. *Distribution*: The distribution of numbers within the system should accurately reflect those used in calculations. Each bit string should be utilised approximately an equal number of times.
2. *Uniqueness*: There should be only one possible bit string for each encoded value.
3. *Generality*: The number system should be defined for bit strings of any length.
4. *Statelessness*: The computation within the number system should be unaffected by any state other than the immediate input data.
5. *Exactness*: Mathematical operations should not introduce errors exceeding the precision of the number system itself. Closure under arithmetic operations is desirable.
6. *Binary Monotonicity*: The mapping of real numbers to bit strings, interpreted as two's complement signed integers, should be monotonic.
7. *Binary Negation*: Negating the bit string as a two's complement signed integer should negate the represented real number and remain invariant for nonnumbers.
8. *Flexibility*: The conversion between bit strings of different lengths should be straightforward.
9. NaR *Propagation*: If an operation yields a NaR (not a real), it should be propagated through all subsequent operations.
10. *Implementation Simplicity*: The number system should be both simple and efficient to implement in hardware, considering factors such as transistor count, energy efficiency, latency, and throughput.

The adoption of a two's complement representation is inherently advantageous for preserving monotonicity and simplifying negation, chiefly because it enables the reutilization of existing components in hardware implementations. In this paper, the GUSTAFSON criteria serve as pivotal benchmarks for assessing a spectrum of number systems.

1.2 Dynamic Range Criteria

Another crucial aspect, particularly related to property 1, is the dynamic range of a given number system – the smallest and largest absolute values it can effectively represent. Irrespective of the chosen dynamic range we propose the following two desirable properties for the dynamic range of a number system:

1a) The largest and smallest positive, as well as the largest and smallest negative representable numbers should be reciprocals of each other.
1b) The dynamic range should be reasonably bounded on both ends as the bit string length approaches infinity.

These properties ensure that increasing the bit-length at a specific point enhances precision exclusively, without unnecessarily extending the dynamic range. An additional implicitly desirable characteristic is the achievement of boundary values even with minimal bit string lengths. However, adhering to property 1a presupposes a highly efficient encoding scheme that rapidly achieves the maximal dynamic range. Consequently, it allows us, without any loss of generality, to focus our discussion primarily on the extent of the upper boundary. Property 1 stipulates that the defined boundary should reflect the magnitudes commonly encountered in computational processes, whilst also accommodating the binary representation of such an upper limit. Should an efficient representation for this upper boundary be ascertained, the determination of the lower boundary will naturally follow in accordance with property 1a.

This aspect can be scrutinised from multiple perspectives, encompassing the practical numbers employed in computations, as well as the efficiency of encoding and the hardware implementation. It is noteworthy at this juncture that for every proposed upper limit, there inevitably exist extreme counterexamples. Nonetheless, the principal aim here is to establish a dynamic range that exhibits overall compatibility with general-purpose computational tasks. While augmenting the dynamic range through an elevation of the upper limit is plausible, it invariably entails a compromise on the coding efficiency for the encompassed numbers. Moreover, it is imperative to delineate between crafting a new format tailored for application-specific representations and devising a novel general-purpose arithmetic number format. The aspiration of this research is to foster the latter.

All number systems examined in this work, inclusive of the one we introduce, adhere to the conventional structure whereby a base is elevated to the power of an exponent. With the base being constant, the dynamic range of the format is delineated by the maximum and minimum values that the exponent can assume. Typically, the exponent is represented as a binary integer, and the base is commonly set to 2. However, within the scope of this paper, we will adopt base $\sqrt{e} \approx 1.649$ for the newly proposed number format and use it as a reference base. The rationale behind selecting base $\sqrt{e}$, in contrast to the traditional base 2, is elaborated upon in Sect. 4.4. Consequently, the discourse concerning the maximum integral exponent is framed in the context of base $\sqrt{e}$, rather than 2.

Certain choices in the binary representation of the maximum exponent offer distinct advantages. When the exponent is represented as a binary number, an ideal feature is the full utilization – or saturation – of the exponent's bit string. Such saturation is evident in exponents like $2^1 - 1$, $2^3 - 1$, $2^7 - 1$, $2^{15} - 1$, $2^{31} - 1$, $2^{63} - 1$, $2^{127} - 1$, $2^{255} - 1$, $2^{511} - 1$, et cetera. Within the framework of a tapered number format, another beneficial property is the efficient encoding of the exponent's bit-length. Therefore, our goal is to ensure that the bit-length of the maximum exponent corresponds to a saturated integer, adhering to the egxponentially growing sequence given by $k \mapsto 2^{2^k} - 1$, which unfolds as $1, 3, 15, 255, 65535$, and beyond.

Upon examining the integers within this sequence, it becomes immediately apparent that the dynamic ranges $\left(\sqrt{e}^{-1}, \sqrt{e}^1 \right) \approx (0.6, 1.6)$, $\left(\sqrt{e}^{-3}, \sqrt{e}^3 \right) \approx (0.2, 4.5)$, and $\left(\sqrt{e}^{-15}, \sqrt{e}^{15} \right) \approx \left(5.5 \times 10^{-4}, 1.8 \times 10^3 \right)$ are insufficiently expansive for general-purpose computational applications. Conversely, the respective dynamic range $\left(\sqrt{e}^{-65535}, \sqrt{e}^{65535}\right) \approx \left(1.8 \times 10^{-14\,231}, 5.6 \times 10^{14\,230}\right)$ is found to be excessively vast, leading to an inefficient allocation of bit representations for numerals that are seldom, if ever, employed in computational tasks. The integer 255 remains as the sole candidate, offering a dynamic range of $(\sqrt{e}^{-255}, \sqrt{e}^{255}) \approx (4.2 \times 10^{-56}, 2.4 \times 10^{55})$. This range holds promise for general-purpose arithmetic, potentially encompassing the spectrum of numbers frequently used in computations without extending into the realm of excessive magnitude. The subsequent analysis aims to ascertain whether this dynamic range aligns with the practical demands of computational applications.

Identifying numbers of significance for computations presents a formidable challenge. One might consider analysing experimental data; however, the selection of representative datasets is inherently difficult due to its subjective nature and the potential for biases specific to certain applications. QUEVEDO historically delineated a desirable dynamic range from $10^{-99} \approx \sqrt{e}^{-658}$ to $10^{99} \approx \sqrt{e}^{658}$ [25, 583], a decision which, albeit practical for maintaining two decimal exponent digits, can be criticised for its pragmatic rather than theoretical basis.

An alternative methodology involves examining attributes that characterise the known universe. These include its age $(4.35 \times 10^{17}\,\text{s})$, diameter $(8.7 \times 10^{26}\,\text{m})$, mass $(8.7 \times 10^{26}\,\text{kg})$, density $(9.9 \times 10^{-27}\,\text{kg}\,\text{m}^{-3})$, along with the cosmological constant (1.1056×10^{-52}) and the HUBBLE constant $(1 \times 10^{-18}\,\text{s}^{-1})$. In addition, dimensionless constants from the standard model – such as the electron neutrino mass (9×10^{-30}), representing the smallest value and thus establishing a lower bound – offer benchmarks for consideration. Extreme values, including the cosmological constant (Λ) at $1.1056 \times 10^{-52}\,\text{m}^{-2}$ and the universe's mass at $1.5 \times 10^{53}\,\text{kg}$, alongside the parameters defining the International System of Units (SI), provide further context. Tables 5 and 6 in Sect. 5.5 offer an illustrative overview of these considerations.

It is also noteworthy that the rescaling of large numbers in computations is a common practice, contingent upon the specific application. For instance, astronomical distances, typically vast, are often expressed in parsecs $(1\,\text{pc} = 3.0857 \times 10^{16}\,\text{m})$, whereas diminutive energy values, prevalent in fields such as atomic, nuclear, and particle physics, are commonly denoted in electronvolts rather than joules $(1\,\text{eV} = 1.602\,176\,634 \times 10^{-19}\,\text{J})$. This rescaling extends into everyday contexts as well; the Kelvin temperature scale, from which the Celsius scale is derived, represents a rescaled unit reflecting the average relative thermal energy of particles within a given gas $(1\,\text{K} = 1.380\,649 \times 10^{-23}\,\text{J})$. This observation underscores the notion that upper bounds derived from physical constants may tend towards the higher end of the spectrum, as various fields and applications routinely rescale numerical values to facilitate practical manipulations within more manageable ranges.

In conclusion, fortuitously, the integer 255 not only demonstrates advantageous bit string attributes but also aligns within the pre-defined bounds for exponents encountered in natural phenomena. Therefore, it is judicious to adopt the boundaries $\sqrt{e}^{-255} \approx 4.2 \times 10^{-56}$ and $\sqrt{e}^{255} \approx 2.4 \times 10^{55}$ as the dynamic range for the discourse within this paper.

2 IEEE 754 Floating-Point Numbers

This paper assumes familiarity with the IEEE 754 floating-point format and directs readers to [8] for a comprehensive introduction, delving into the complexities inherent in the multitude of special cases and intricacies intrinsic to the standard. To aid the reader, Table 1 furnishes the fixed number of exponent and fraction bits, denoted as n_e and n_f respectively, alongside the dynamic range corresponding to each admissible bit string length. Additionally, this table encapsulates recent proprietary parameterizations that will serve as benchmarks for subsequent comparisons (Fig. 1).

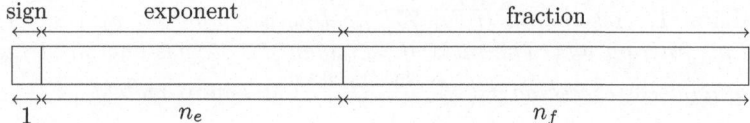

Fig. 1. IEEE 754 floating-point number bit string format.

Table 1. Overview of different IEEE 754 floating-point formats, their dynamic ranges and the ratio of bit strings that are redundant or exceeding the dynamic range $\sqrt{e}^{-255} \approx 4.2 \times 10^{-56}$ to $\sqrt{e}^{255} \approx 2.4 \times 10^{55}$ ('waste').

name	n	n_e	n_f	smallest	largest	waste/%	notes
float8	8	4	3	$\approx 2.0 \times 10^{-3}$	$\approx 2.4 \times 10^{2}$	5.08	not standardised
float16	16	5	10	$\approx 6.0 \times 10^{-8}$	$\approx 6.6 \times 10^{4}$	3.12	
bfloat16	16	8	7	$\approx 1.2 \times 10^{-38}$	$\approx 3.4 \times 10^{38}$	0.78	[26], no subnormals
TF32	19	8	10	$\approx 1.2 \times 10^{-38}$	$\approx 3.4 \times 10^{38}$	0.78	[17], no subnormals
float32	32	8	23	$\approx 1.4 \times 10^{-45}$	$\approx 3.4 \times 10^{38}$	0.39	
float64	64	11	52	$\approx 4.9 \times 10^{-324}$	$\approx 1.8 \times 10^{308}$	82.03	
float128	128	15	112	$\approx 6.5 \times 10^{-4966}$	$\approx 1.2 \times 10^{4932}$	98.88	
float256	256	19	236	$\approx 2.2 \times 10^{-78\,984}$	$\approx 1.6 \times 10^{78\,913}$	99.93	

The proprietary bfloat16 (brain float) and TF32 (TensorFloat-32) formats were developed as alternatives to the float16 format, which exhibits insufficient dynamic range due to its small exponent. This limitation has become especially prominent in machine learning applications in recent years [17]. Notably, these formats omit subnormal numbers, implying that any number smaller than the smallest normal number is simply rounded to zero. The rationale behind this omission is the costliness of implementing subnormal numbers, which is deemed unwarranted for the minimal gain in dynamic range. However, it's essential to acknowledge that this choice deviates from the IEEE 754 floating-point standard.

An additional metric presented in Table 1 is the ratio of 'wasted' bit strings. This metric encompasses not only redundant NaN or unused subnormal representations but also all binary representations of values exceeding the previously determined target dynamic range from $\sqrt{e}^{-255} \approx 4.2 \times 10^{-56}$ to $\sqrt{e}^{255} \approx 2.4 \times 10^{55}$. It is important to note that the metric of 'waste' employed in this paper serves as a quantitative measure to elucidate the inefficiency of a number system that offers any excessively high dynamic range. This serves to illustrate the tendency for human intuition to underestimate the drawbacks of such designs and combinatoric effects. It is not our intention to enshrine the previously determined dynamic range $\sqrt{e}^{-255}$ to $\sqrt{e}^{255}$ as an immutable benchmark, but rather to present it as one potential reference point. It holds

Proposition 1. *Assume an IEEE 754 floating-point format with n_e exponent bits and n_f fraction bits. The ratio of redundant bit strings and bit strings representing numbers exceeding $\pm \left(\sqrt{e}^{-255}, \sqrt{e}^{255} \right)$ is approximately*

$$[1 + (n_e \geq 10) \cdot (2^{n_e} - 734) + (n_e \geq 10 \vee \text{no subnormals})] \cdot 2^{-n_e} - 3 \cdot 2^{1-n_e-n_f}. \quad (1)$$

Proof. See Section B.

Please refer to Table 1 for an overview of values for the different IEEE 754 and proprietary floating-point types. Concerning the earlier mentioned GUSTAFSON criteria, it can be argued that IEEE 754 floating-point numbers fail to satisfy any of them [10]. Notably, `float16` and `float32` exhibit inadequate dynamic range relative to the desired $\sqrt{e}^{-255} \approx 4.2 \times 10^{-56}$ and $\sqrt{e}^{255} \approx 2.4 \times 10^{55}$. Conversely, `float64` and larger types exhibit an excessive dynamic range, violating property 1. The resultant ratio of wasted bit strings due to excessive dynamic range, occurring only for $n_e \geq 9$, leads to a significant number of unused bit strings in the respective IEEE 754 floating-point formats. For instance, double-precision floating-point numbers (`float64`) allocate approximately 82% of their available bit strings to representations that are unlikely to be used in calculations, violating properties 1a and 1b. This becomes even more profound for `float128` and `float256`, where approximately 99% and almost 100% are respectively wasted.

Beyond issues of dynamic range, the presence of numerous redundant NaN representations directly contravenes property 2. This issue particularly impacts smaller IEEE 754 floating-point formats, leading to a significant waste of bit patterns – 5.08% and 3.12% in `float8` and `float16`, respectively. The failure to satisfy properties 6 and 7 introduces a notable overhead in hardware. Concerning NaR propagation, as illustrated by examples such as `pow(NaN, 0) = 1` and `pow(1, NaN) = 1` for 'quiet' NaN's, there are violations of property 9. Further discussion on these issues is available in [10].

3 Posits

Since their introduction by GUSTAFSON and YONEMOTO in 2017 [10], posits, as an alternative number system to IEEE 754 floating-point numbers, have undergone extensive analysis and evolved into a standard draft [11]. Posits now represent the current state of the art for tapered floating-point formats, which go back to MORRIS [22], and are used in a wide range of fields [14,24]. The standard draft [11] defines posits as follows:

Definition 1 (posit encoding). *Let $n \in \mathbb{N}$ with $n \geq 5$. Any n-bit MSB→LSB string of the form*

with

$$S \qquad\qquad : sign\ bit \tag{2}$$

$$R := (R_{k-1}, \ldots, R_0) \qquad\qquad : regime\ bits \tag{3}$$

$$\overline{R_0} \qquad\qquad : regime\ termination\ bit \tag{4}$$

$$r := \begin{cases} -k & R_0 = 0 \\ k-1 & R_0 = 1 \end{cases} \qquad\qquad : regime \tag{5}$$

$$E := (E_1, E_0) \qquad\qquad : exponent\ bits \tag{6}$$

$$e := 2E_1 + E_0 \qquad\qquad : exponent \tag{7}$$

$$p := n - k - 4 \in \{0, \ldots, n-5\} \qquad\qquad : fraction\ bit\ count \tag{8}$$

$$F := (F_{p-1}, \ldots, F_0) \in \{0,1\}^p \qquad\qquad : fraction\ bits \tag{9}$$

$$f := 2^{-p} \sum_{i=0}^{p-1} F_i 2^i \in [0,1) \qquad\qquad : fraction \tag{10}$$

$$\hat{e} := (-1)^S (4r + e + S) \qquad\qquad : `actual'\ exponent \tag{11}$$

encodes the posit value

$$\pi((S, R, \overline{R_0}, E, F)) := \begin{cases} 0 & S=0 \\ \mathrm{NaR} & S=1 \\ [(1-3S)+f] \cdot 2^{\hat{e}} & otherwise. \end{cases} \quad R = \overline{R_0} = E = F = \mathbf{0} \tag{12}$$

with $\pi \colon \{0,1\}^n \mapsto \{0, \mathrm{NaR}\} \cup \pm [2^{-4n+8}, 2^{4n-8}]$. Without loss of generality, any bit string shorter than 5 bits is also included in the definition by assuming the missing bits to be zero bits ('ghost bits'). The colour scheme for the different bit string segments was adopted from the standard [11].

Posits were explicitly designed to satisfy all GUSTAFSON criteria (refer to [8]). However, there has been a modification in the standard draft [11] to further accommodate property 8 after the publication of [10]: The number of exponent bits was fixed at 2, a departure from the previous variation depending on n. While a fixed exponent size facilitates easy conversion between different precisions, it introduces a trade-off between precision and dynamic range.

Another alteration since [10] is the replacement of ∞ with NaR. This change was made in favour of property 9, enabling the propagation of nonreal values instead of termination, albeit at the cost of allowing division by zero and ∞. While valid arguments support the inclusion of either NaR or ∞, the underlying universal wheel algebra [4] is defined with both elements (a bottom element $\bot := 0/0$ and infinity ∞). However, including both elements is impractical as it would compromise the symmetry of the model. The handling conventions for NaR are further discussed in Sect. 4.6.

We will now examine the distribution of numbers within posits' dynamic range, which spans from 2^{-4n+8} to 2^{4n-8} (see Table 2 for an overview). Of

Table 2. Overview of different posit precisions and their dynamic ranges.

name	n	smallest	largest
posit8	8	$\approx 5.96 \times 10^{-8}$	$\approx 1.68 \times 10^{7}$
posit16	16	$\approx 1.39 \times 10^{-17}$	$\approx 7.21 \times 10^{16}$
posit32	32	$\approx 7.52 \times 10^{-37}$	$\approx 1.32 \times 10^{36}$
posit64	64	$\approx 2.21 \times 10^{-75}$	$\approx 4.52 \times 10^{74}$
posit128	128	$\approx 1.91 \times 10^{-152}$	$\approx 5.24 \times 10^{151}$
posit256	256	$\approx 1.42 \times 10^{-306}$	$\approx 7.02 \times 10^{305}$

particular interest is the ratio of bit strings that exceed the targeted dynamic range as previously discussed. In Proposition 1, we discovered that `float64`, and formats with greater precision, allocate a considerable proportion of bit strings to superfluous numerical representations. In the case of posits, we can show the following

Proposition 2. *Let $n \in \mathbb{N}_1$. The ratio of posit bit strings of length n representing numbers exceeding $\pm \left(\sqrt{e}^{-255}, \sqrt{e}^{255} \right)$ is approximately*

$$\begin{cases} 0 & n \leq 47 \\ \frac{4 \cdot 2^{n-48}}{2^n} & n \geq 48 \end{cases} = \begin{cases} 0 & n \leq 47 \\ 2^{-46} & n \geq 48 \end{cases} \approx \begin{cases} 0 & n \leq 47 \\ 1.42 \times 10^{-14} & n \geq 48. \end{cases} \tag{13}$$

Proof. See Section C.

Remarkably, the ratio of excessive bit patterns in relation to dynamic range remains consistently small across all n. It is important to note, however, that this argument pertains solely to the dynamic range and does not assess the efficiency of the posit format itself.

Overall, alongside meticulous design choices aimed at mitigating the inherent redundancy in IEEE 754 floating-point numbers, the primary quantitative disparity between them and posits resides in the variable-length exponent of posits. This characteristic affords heightened precision for values with exponents proximate to zero, which, in practical terms, constitute the numbers predominantly employed in computational tasks. However, this advantage entails a corresponding trade-off, leading to diminished precision for values characterized by large exponents. This trade-off is further elucidated in [8,10].

We have demonstrated that the variable-length exponent not only enhances the precision of posits for commonly encountered values but also protects them from excessive bit allocation for extremely large numbers. This stands in stark contrast to IEEE 754 floating-point numbers, which squander a significant number of bits (refer to Table 1). However, it is important to note that the dynamic range of posits is relatively limited for bit string lengths most relevant below 64. Furthermore, posits encode exponents of substantial magnitude relatively inefficiently due to the necessity of lengthy regimes, resulting in a scarcity of bit strings allocated for numbers with large-magnitude exponents, which puts the results of Proposition 2 into perspective.

If one were to optimise the exponent encoding of posits, it would result in a number format boasting a significantly expanded dynamic range and a greater abundance of available bit strings for each exponent. However, a concomitant increase in the ratio of unused bit strings would ensue. This phenomenon is exemplified in Sect. 4 and underpins the adoption of a constrained dynamic range approach for the number system delineated in this study.

4   Takum Arithmetic

The preceding sections have introduced both IEEE 754 floating-point numbers and posits, offering an extensive discourse on their respective strengths and weaknesses. Within the realm of posits and by design, the exponent coding emerges as an area ripe for enhancement, given that the sign and fraction bits exhibit maximal information density/entropy in both formats.

Regarding posit exponent coding, the utilisation of a prefix code to delineate regimes engenders sequences of low-entropy runs (sequences of consecutive ones or zeros which can be interpreted to have low information content), presenting an opportunity for optimisation. While alternative universal codes have been subjected to rigorous scrutiny [20], they entail a significant overhead, particularly evident for small exponents (see the later discussion in Sect. 5.3).

A hypothetical strategy for optimizing the posit exponent encoding entails defining k as the number of exponent bits minus 2 and supplementing the prefix code with a variable bitwise representation of the exponent. An implicit most significant bit (MSB) of 1 is presumed for $k > 1$. While this approach yields coding efficiency akin to that of posits for diminutive exponents, its adoption engenders a considerable expansion in the dynamic range. For instance, with a 16-bit configuration, the largest number, $01\ldots1$, attains $k = 15$. Consequently, the exponent spans a bit-length of 17 (inclusive of the implicit 1 bit), manifesting in the binary string 10000000000000000. This corresponds to an excessively large exponent of 65536.

This expansion results in an abundance of redundant bit strings for numbers that significantly exceed the intended dynamic range, thereby rendering such bit strings superfluous. Attempts to address this issue by imposing constraints on the total length of regime and exponent bits disrupt the symmetry of the dynamic range, a characteristic deemed undesirable according to property 1a. The pivotal observation is that the coding of exponents must be *intrinsically* bounded instead. Consequently, there should come a point where additional bits appended to the bit string contribute solely to precision rather than dynamic range, as they cease to impact the exponent coding.

4.1   Definition

Let us consider a radical question: If we constrain the dynamic range of the exponent, is there even a necessity for prefix codes to encode exponents of arbitrary magnitude? Previously, we elucidated a target dynamic range spanning from $\sqrt{e}^{-255}$ to $\sqrt{e}^{255}$, underscoring the significance of the binary representation of 255 as having a length of 8, a power of 2. Given that the leading bit of any non-zero number is 1, we can effectively encode the bit-length of any number ranging from 1 to 255 using merely three 'regime' bits (capable of expressing any regime value between 0 and 7 for the number of 'explicit' bits following the implicit one bit). By appending the explicit bits to the regime bits, we achieve a highly efficient variable-length representation for numbers larger or equal to one.

To also encode zero, we subtract 1 from the value, yielding an encoding for numbers within the range 0 to 254. It may raise the question whether this approach compromises our intended dynamic range. However, in reality, it enables us to precisely match our target dynamic range from the outset. Since the significand of a base-$\sqrt{e}$ floating-point representation falls within the interval $[1, \sqrt{e})$, our aim is to confine the exponents within the range -255 to 254, corresponding to a dynamic range of $\sqrt{e}^{-255}$ to $\sqrt{e}^{255}$. Let us designate the three regime bits in mauve and the explicit bits in blue. The values from 0 to 8 are encoded as follows: 000, 0010, 0011, 01000, 01001, 01010, 01011, and 011000. The value 254 is encoded as 1111111111 (10 bits), significantly shorter than the posit encoding requiring 68 bits.

One advantage of the posits' prefix codes is the ability to store an additional bit R_0 of information, depending on whether the regime consists of all zeros terminated by a one bit or all ones terminated by a zero bit, implying $R_0 = 0$ and $R_0 = 1$ respectively. However, for encoding a complete exponent with our scheme, we require an additional 'direction' bit instead to indicate when to apply a bias. Based on the aforementioned encoding scheme, we propose the following format:

Definition 2 (takum encoding). *Let* $n \in \mathbb{N}$ *with* $n \geq 12$. *Any* n-*bit* $MSB{\to}LSB$ *string* $(S, D, R, C, M) \in \{0,1\}^n$ *of the form*

$$
\begin{array}{|c|c|c|c|c|}
\hline
S & D & R & C & M \\
\hline
\end{array}
$$

$$
\underset{1}{\underbrace{}}\ \underset{1}{\underbrace{}}\ \underset{3}{\underbrace{}}\ \underset{r}{\underbrace{}}\ \underset{p}{\underbrace{}}
$$

with

$$S \in \{0,1\} \qquad\qquad : sign\ bit \qquad\qquad (14)$$

$$D \in \{0,1\} \qquad\qquad : direction\ bit \qquad (15)$$

$$R := (R_2, R_1, R_0) \in \{0,1\}^3 \qquad\qquad : regime\ bits \qquad (16)$$

$$r := \begin{cases} 7 - (4R_2 + 2R_1 + R_0) & D = 0 \\ 4R_2 + 2R_1 + R_0 & D = 1 \end{cases} \in \{0,\dots,7\} \quad : regime \qquad (17)$$

$$C := (C_{r-1}, \dots, C_0) \in \{0,1\}^r \qquad\qquad : characteristic\ bits \quad (18)$$

$$c := \begin{cases} -2^{r+1} + 1 + \sum_{i=0}^{r-1} C_i 2^i & D = 0 \\ 2^r - 1 + \sum_{i=0}^{r-1} C_i 2^i & D = 1 \end{cases} \qquad : characteristic \qquad (19)$$

$$p := n - r - 5 \in \{n - 12, \dots, n - 5\} \qquad : mantissa\ bit\ count \quad (20)$$

$$M := (M_{p-1}, \dots, M_0) \in \{0,1\}^p \qquad\qquad : mantissa\ bits \qquad (21)$$

$$m := 2^{-p} \sum_{i=0}^{p-1} M_i 2^i \in [0,1) \qquad\qquad : mantissa \qquad (22)$$

$$\ell := (-1)^S (c + m) \in (-255, 255) \qquad : logarithmic\ value \quad (23)$$

encodes the takum value

$$
\tau((S, D, R, C, M)) := \begin{cases} \begin{cases} 0 & S = 0 \\ \mathrm{NaR} & S = 1 \end{cases} & D = R = C = M = \mathbf{0} \\ (-1)^S \sqrt{\mathrm{e}^\ell} & otherwise \end{cases} \qquad (24)
$$

with $\tau \colon \{0,1\}^n \mapsto \{0, \mathrm{NaR}\} \cup \pm \left(\sqrt{\mathrm{e}^{-255}}, \sqrt{\mathrm{e}^{255}} \right)$ *and* EULER*'s number* e. *Without loss of generality, any bit string shorter than 12 bits is also considered in the definition by assuming the missing bits to be zero bits ('ghost bits').*

The term 'takum' originates from the Icelandic phrase 'takmarkað umfang', translating to 'limited range'. Pronounced initially akin to the English term 'tug', with a shortened 'u' sound, the 'g' is articulated as a 'k'. The '-um' follows swiftly after the 'k', pronounced akin to 'um' in the English term 'umlaut'.

This format specification initiates a shift in nomenclature, substituting the exponent bits E and exponent e with characteristic bits C and characteristic c. This modification originates from the intrinsic logarithmic characteristics of takum arithmetic, as elaborated upon in Sect. 4.3. Such a foundation more aptly embodies the substitution of the traditional integral exponent with a logarithmic value, ℓ, which consists of a characteristic, c – the former representing the integral portion, and the latter, a mantissa, m, delineating the fractional portion. In comparison to the posit definition (refer to Definition 1), this adjustment notably simplifies (12).

Additionally, we introduce a takum colour scheme, prioritising uniformity in both lightness and chroma within the perceptually uniform OKLCH colour space [19]. Detailed colour definitions are delineated in Table 3.

Table 3. Overview of the takum arithmetic colour scheme.

colour	identifier	OKLCH	CIELab	HEX (sRGB)
	sign	$(50\%, 0.17, 25)$	$(40.28, 53.78, 33.31)$	#B02A2D
	direction	$(50\%, 0.17, 142.5)$	$(43.91, -45.71, 46.68)$	#007900
	regime	$(50\%, 0.17, 320)$	$(39.29, 46.60, -39.19)$	#8C399E
	characteristic	$(50\%, 0.17, 260)$	$(40.34, 10.54, -59.37)$	#1F5DC2
	mantissa	$(50\%, 0.00, 0)$	$(42.00, 0.00, 0.00)$	#636363

We observe that the additional information previously represented by R_0 is now integrated into the 'direction bit' D. The designation has been chosen due to the bit's function in indicating whether the logarithmic value ℓ increases $(D = 1)$ or decreases $(D = 0)$ with the incrementation of the takum bit string. In other words, the bit signifies whether the growth of the logarithmic value is aligned with the growth of the number, which can also be referred to as both pointing in the same direction. It is noteworthy that, in general, the total length of the characteristic bit string segment $(1 + 3 + r)$ never exceeds 11 bits, and that we precisely align with the targeted dynamic range of $\sqrt{e}^{-255}$ to $\sqrt{e}^{255}$. For a succinct elucidation of the takum encoding scheme, please refer to Table 4, which presents a small selection of examples.

4.2 Rounding

In terms of rounding, we adhere to the posit standard [11, Section 4.1] by employing 'saturation arithmetic'. This approach ensures that there is no under- or overflow for numbers outside the dynamic range; instead, they are clamped to the smallest or largest representable number, respectively. Saturation arithmetic finds justification in the fact that the error introduced by saturation is consistently smaller than the potentially infinite error resulting from overflow, and it eliminates scenarios where a number of infinitesimally small magnitude vanishes.

Table 4. Examples illustrating the takum encoding scheme.

bits	r	c	m	ℓ	t
01	0	0	0	$+(0+0) = 0$	$+\sqrt{e}^0 = 1$
01000001	0	0	0.125	$+(0+0.125) = 0.125$	$+\sqrt{e}^{0.125} \approx 1.1$
11	0	0	0	$-(0+0) = 0$	$-\sqrt{e}^0 = -1$
11000001	0	0	0.125	$-(0+0.125) = -0.125$	$-\sqrt{e}^{-0.125} \approx -0.9$
01001	1	1	0	$+(1+0) = 1$	$+\sqrt{e}$
001	3	-15	0	$+(-15+0) = -15$	$+\sqrt{e}^{-15} \approx 5.5e-4$
010001	0	0	0.5	$+(0+0.5) = 0.5$	$+\sqrt{e}^{0.5} \approx 1.3$
101111	0	-1	0.5	$-(-1+0.5) = 0.5$	$-\sqrt{e}^{0.5} = -1.3$
10010111111	5	-32	0.5	$-(-32+0.5) = 31.5$	$-\sqrt{e}^{31.5} \approx -6.9e6$
10011000000	4	-31	0	$-(-31+0) = 31$	$-\sqrt{e}^{31} \approx -5.4e6$
10011000001	4	-31	0.25	$-(-31+0.25) = 30.75$	$-\sqrt{e}^{30.75} \approx -4.8e6$
1000000000001	7	-255	0.5	$-(-255+0.5) = 254.5$	$-\sqrt{e}^{254.5} \approx -1.8e55$
1111111111111	7	254	0.5	$-(254+0.5) = -254.5$	$-\sqrt{e}^{-254.5} \approx -5.4 \times 10^{-56}$
0000000000001	7	-255	0.5	$+(-255+0.5) = -254.5$	$+\sqrt{e}^{-254.5} \approx 5.4e-56$
0111111111111	7	254	0.5	$+(254+0.5) = 254.5$	$+\sqrt{e}^{254.5} \approx 1.8e55$

The significantly expanded dynamic range offered by takums confers a notable advantage in that saturation cases occur much less frequently compared to posits. Refer to Algorithm 1 for the rounding procedure, which initially clamps values outside the dynamic range. For values falling within the dynamic range, the procedure first converts them to $n+1$ bit truncated takums and then performs rounding based on the least significant bit (LSB). Under- and overflows are subsequently corrected to 0 and NaR, respectively, as the final step.

As evident, rounding takes place within the logarithmic domain, where the logarithmic value ℓ is rounded – a conventional practice in logarithmic number systems. The rounding midpoint between two numbers corresponds to their geometric mean rather than their arithmetic mean. Notably, this method of rounding remains consistent with roundings in the non-mantissa bits. In contrast, posits, with their linear significant, inherently employ two types of roundings: geometric mean rounding in the non-fraction bits and arithmetic mean rounding in the fraction bits.

While this impedes the formal analysis of posits in low-precision applications, it presents an opportunity for future research to develop a comprehensive theory of takum rounding encompassing both mantissa and non-mantissa bits, extending beyond the scope of Proposition 11. Such an endeavour holds considerable significance for low-precision applications, particularly those where representations predominantly feature zero mantissa bits and rounding predominantly affects the non-mantissa bits.

Algorithm 1: Takum rounding algorithm yielding $\mathrm{round}_n(x)$ for a number $x \in \mathbb{R} \cup \{\mathrm{NaR}\}$ to $n \in \mathbb{N}_2$ bits. The lossless takum encoding function τ^{inv} is defined in Proposition 8, the $\mathrm{truncate}_i$ function zero-extends or strips off LSBs until the bit string has the desired length $i \in \mathbb{N}_0$.

---

input : $x \in \mathbb{R} \cup \{\mathrm{NaR}\}$: input
    $n \in \mathbb{N}_1$: bit count
output: $\mathrm{round}_n(x) \in$
    $\{0, \mathrm{NaR}\} \cup \pm\left(\sqrt{e}^{-255}, \sqrt{e}^{255}\right)$

/* saturate excessive numbers */
if $x \in \left(-\infty, -\sqrt{e}^{255}\right]$ **then**
$\left| \quad T \leftarrow \begin{cases} (1,1) & n = 2 \\ (1,0,1) & n \geq 3 \end{cases} \in \{0,1\}^n \right.$
else if $x \in \left[-\sqrt{e}^{-255}, 0\right)$ **then**
$\left| \quad T \leftarrow 1 \in \{0,1\}^n \right.$
else if $x \in \left(0, \sqrt{e}^{-255}\right]$ **then**
$\left| \quad T \leftarrow (0,1) \in \{0,1\}^n \right.$
else if $x \in \left[\sqrt{e}^{255}, \infty\right)$ **then**
$\left| \quad T \leftarrow (0,1) \in \{0,1\}^n \right.$

else
$\left|\right.$ $T \leftarrow \mathrm{truncate}_{n+1}(\tau^{\mathrm{inv}}(x))$
$\left|\right.$ /* round */
$\left|\right.$ **if** $T_0 = 0$ **then**
$\left|\right.$ $\left| \quad T \leftarrow \mathrm{truncate}_n(T) \right.$
$\left|\right.$ **else**
$\left|\right.$ $\left| \quad T \leftarrow \mathrm{truncate}_n(T) + 1 \right.$
$\left|\right.$ **end**
$\left|\right.$ /* saturate over-/underflows */
$\left|\right.$ **if** $T = 0 \wedge x \neq 0$ **then**
$\left|\right.$ $\left| \quad T \leftarrow T + \mathrm{sign}(x) \right.$
$\left|\right.$ **else if** $T = (1,0) \wedge x \neq \mathrm{NaR}$ **then**
$\left|\right.$ $\left| \quad T \leftarrow T - \mathrm{sign}(x) \right.$
$\left|\right.$ **end**
end
$\mathrm{round}_n(x) \leftarrow \tau(T)$

4.3   Logarithmic Significand

Besides the encoding scheme, takums also diverge from posits due to their logarithmic significand. This choice stems from promising outcomes observed in the application of logarithmic significands to posits and the promising qualities of logarithmic number systems in general [21,24]. Although the result provided in Definition 2 appears straightforward, it conceals the underlying derivation. This section aims to elucidate this derivation and expound upon the advantages of a logarithmic number system. As formalised in [20, (1)] and extended here for base $\sqrt{e}$, any real number $x \in \mathbb{R} \setminus \{0\}$ can be represented as a floating-point number

$$(-1)^s \times \sqrt{e}^h \times \sigma(f), \tag{25}$$

where $s \in \{0,1\}$ denotes the sign, $h \in \mathbb{Z}$ signifies the exponent, $f \in [0,1)$ denotes the fraction, and the mapping $\sigma: [0,1) \mapsto [1, \sqrt{e})$ represents the significand. The linear significand $\sigma(f) = 1 + (\sqrt{e} - 1)f$ is the conventional choice (for base 2 it would be the function $f \mapsto 1 + f$). While this representation may seem unconventional, it is imperative to recognise that the base 2 for floating-point arithmetic is, in theory, not immutable. We possess the liberty to conceptualise alternative representations, temporarily setting aside considerations pertaining to implementation efficiency.

As an alternative to the linear significand, a base-$\sqrt{e}$ logarithmic number system utilises the logarithmic significand $\sigma(f) = \sqrt{e}^f$, facilitating the representation of x as $(-1)^s \times \sqrt{e}^{h+f} =: (-1)^s \times \sqrt{e}^{\ell}$, where $\ell \in \mathbb{R}$ signifies the logarithmic value of x. There exists considerable confusion surrounding the terminology – 'exponent', 'fraction', 'characteristic', and 'mantissa' – within the context of floating-point numbers. The terms 'characteristic' and 'mantissa' originate from logarithm tables, delineating the integral and fractional components of a logarithm, respectively. In 1946, BURKS et al. (as published in [3]) employed this nomenclature to denote the exponent and fraction of a floating-point number. Nevertheless, given that floating-point numbers do not conform strictly to logarithmic principles, this terminology is considered inaccurate [16], a stance reflected in the absence of these terms in the current IEEE 754 standard [12]. For the logarithmic significand, it is appropriate to revert the terminology, naming the exponent as 'characteristic' $c := h$ and the fraction as 'mantissa' $m := f$, a convention also closely adopted in the definition of takums. A comparative illustration of both significands is presented in Fig. 2.

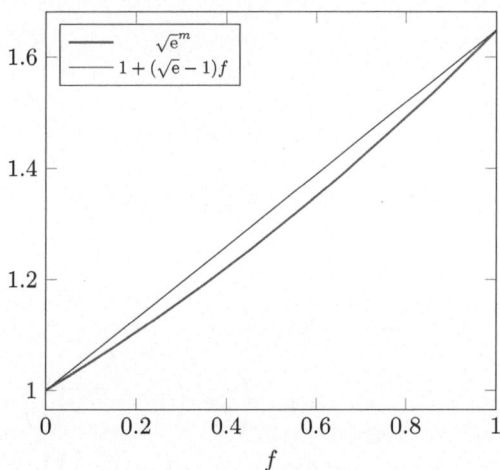

Fig. 2. A comparison of the linear and logarithmic significands.

Arithmetic operations such as multiplication, division, inversion, square root, and squaring on numbers of this form become remarkably straightforward. Given $x, \tilde{x} \in \mathbb{R} \setminus \{0\}$, where $x = (-1)^s \sqrt{e}^{\ell}$ and $\tilde{x} = (-1)^{\tilde{s}} \sqrt{e}^{\tilde{\ell}}$, with $s, \tilde{s} \in \{0,1\}$ and $\ell, \tilde{\ell} \in \mathbb{R}$, it is observed that:

$$x \cdot \tilde{x} = (-1)^s \sqrt{e}^{\ell} \cdot (-1)^{\tilde{s}} \sqrt{e}^{\tilde{\ell}} = (-1)^{s+\tilde{s}} \sqrt{e}^{\ell+\tilde{\ell}}, \tag{26}$$

$$x \div \tilde{x} = (-1)^s \sqrt{e}^{\ell} \div (-1)^{\tilde{s}} \sqrt{e}^{\tilde{\ell}} = (-1)^{s-\tilde{s}} \sqrt{e}^{\ell-\tilde{\ell}}, \tag{27}$$

$$x^{-1} = \left((-1)^s \sqrt{e}^{\ell} \right)^{-1} = (-1)^s \sqrt{e}^{-\ell}, \tag{28}$$

$$\sqrt{|x|} = \sqrt{\sqrt{e}^{\ell}} = \sqrt{e}^{\frac{\ell}{2}}, \tag{29}$$

$$x^2 = \left((-1)^s \sqrt{e}^{\ell} \right)^2 = \sqrt{e}^{2\ell}. \tag{30}$$

If ℓ and $\tilde{\ell}$ are stored as fixed-point numbers, these operations reduce to fixed-point additions, subtractions, negations, and bit-shifts, all of which are highly efficient. The primary challenge lies in additions and subtractions: Without loss of generality, assuming $s = \tilde{s} = 0$, $\ell > \tilde{\ell}$ and $q := \ell - \tilde{\ell} > 0$

$$\log_{\sqrt{e}}(x + \tilde{x}) = \log_{\sqrt{e}}\left(\sqrt{e}^{\ell} + \sqrt{e}^{\tilde{\ell}} \right) \tag{31}$$

$$= \log_{\sqrt{e}}\left(\sqrt{e}^{\ell} \left(1 + \sqrt{e}^{\tilde{\ell}-\ell} \right) \right) \tag{32}$$

$$= \log_{\sqrt{e}}\left(\sqrt{e}^{\ell} \left(1 + \sqrt{e}^{-q} \right) \right) \tag{33}$$

$$= \ell + \log_{\sqrt{e}}\left(1 + \sqrt{e}^{-q} \right) \tag{34}$$

and

$$\log_{\sqrt{e}}(x - \tilde{x}) = \log_{\sqrt{e}}\left(\sqrt{e}^{\ell} - \sqrt{e}^{\tilde{\ell}} \right) \tag{35}$$

$$= \log_{\sqrt{e}}\left(\sqrt{e}^{\ell} \left(1 - \sqrt{e}^{\tilde{\ell}-\ell} \right) \right) \tag{36}$$

$$= \log_{\sqrt{e}}\left(\sqrt{e}^{\ell} \left(1 - \sqrt{e}^{-q} \right) \right) \tag{37}$$

$$= \ell + \log_{\sqrt{e}}\left(1 - \sqrt{e}^{-q} \right). \tag{38}$$

In essence, to determine the exponent resulting from the addition or subtraction of x and $\tilde{x}$, one adds $\log_{\sqrt{e}}\left(1 + \sqrt{e}^{-q} \right)$ or $\log_{\sqrt{e}}\left(1 - \sqrt{e}^{-q} \right)$ to x respectively. These logarithmic calculations are commonly referred to as GAUSSian logarithms and can be defined as

$$\Phi_b^+(q) := \log_b(1 + b^{-q}), \tag{39}$$

$$\Phi_b^-(q) := \log_b(1 - b^{-q}) \tag{40}$$

for $q > 0$ and a general base $b > 1$. Efficient hardware implementations, with up to 32 bits precision, have been demonstrated using lookup tables and interpolation [2,5,6]. These implementations offer comparable, if not superior, latency to addition and subtraction using linear significands, alongside significantly enhanced overall arithmetic performance and reduced power consumption. For an exploration of the handling of even greater levels of precision, where tables don't scale well anymore, the reader is directed to Sect. 4.4. This section further elucidates why this is also the rationale behind our selection of the base $\sqrt{e}$.

A significant advantage of logarithmic number systems overall lies in the singular focus required within FPU design, as opposed to linear significands which necessitate dedicated logic design and optimisation for operations like multiplication, division, inversion, square roots, and powers. Posits employing logarithmic significands have previously been explored and found to be well-suited for neural networks [14]. Notably, posits and takums with logarithmic significands offer the advantageous feature of perfect invertibility for every number, aligning closely with the original concept of unums [9]. Furthermore, this characteristic simplifies hardware implementations as inversion can be accomplished through a bitwise operation (see Proposition 7).

All of these considerations prompted the default definition of takums with a logarithmic significand and the format is so tailored for being a logarithmic number system that defining it with a linear significand makes little sense, mostly because of the base $\sqrt{e}$ that cannot properly translate into bit-shifts for arithmetic. If we do it anyway, (23) and (24) would change to

$$\bar{\ell} = (-1)^S (c + S) \qquad (41)$$

and

$$\bar{t}((S, D, R, C, M)) :=$$

$$\begin{cases} \begin{cases} 0 & S = 0 \\ \mathrm{NaR} & S = 1 \end{cases} & D = R = C = M = 0 \\ [(1 - (1 + \sqrt{e})S) + (\sqrt{e} - 1)f] \cdot \sqrt{e}^{\bar{\ell}} & \text{otherwise.} \end{cases} \qquad (42)$$

We observe that the fraction $f \in [0, 1)$ is linearly mapped to a value in the interval $[1, \sqrt{e})$ when $S = 0$, and to the interval $[-\sqrt{e}, -1)$ when $S = 1$, through the mapping

$$f \mapsto \begin{cases} 1 + (\sqrt{e} - 1)f & S = 0 \\ -\sqrt{e} + (\sqrt{e} - 1)f & S = 1 \end{cases} = (1 - (1 + \sqrt{e})S) + (\sqrt{e} - 1)f. \qquad (43)$$

We can insert the logarithmic significand

$$m \mapsto \begin{cases} \sqrt{e}^m & S = 0 \\ -\sqrt{e}^{1-m} & S = 1 \end{cases} = (-\sqrt{e})^S \sqrt{e}^{(-1)^S m} \qquad (44)$$

and (41) into (42) to derive (24) as follows

$$(-\sqrt{e})^S \sqrt{e}^{(-1)^S m} \sqrt{e}^{\overline{\ell}} = (-\sqrt{e})^S \sqrt{e}^{(-1)^S m} \sqrt{e}^{(-1)^S (c+S)} \tag{45}$$

$$= (-\sqrt{e})^S \sqrt{e}^{(-1)^S (c+m+S)} \tag{46}$$

$$= (-1)^S \sqrt{e}^S \sqrt{e}^{(-1)^S (c+m+S)} \tag{47}$$

$$= (-1)^S \sqrt{e}^{(-1)^S (c+m+S)+S} \tag{48}$$

$$= (-1)^S \sqrt{e}^{(-1)^S (c+m)} \tag{49}$$

$$= (-1)^S \sqrt{e}^{\ell}, \tag{50}$$

with ℓ as in (23). As evident from the S-addition present in the exponent formula (41), which is analogous to the original posit exponent (refer to (11) in Definition 1), a notable cancellation occurs, resulting in a remarkably straightforward expression for the takum value. This observation also lends support to the adoption of a logarithmic significand, as it not only simplifies the mathematical formulation but also facilitates subsequent formal analyses, as demonstrated in Sect. 4.5.

4.4   Choice of Base $\sqrt{e}$

Another crucial aspect warranting discussion is the selection of $\sqrt{e}$ as the base of the takum logarithmic number system, namely in the expression $(-1)^S \sqrt{e}^{\ell}$ in (24), where much more common alternatives such as $\sqrt{2}$, 2 or $\sqrt[3]{2}$ could have been chosen.

While floating-point numbers with linear significands necessitate a base of 2 to facilitate significand shifts for arithmetic, no such restriction applies to logarithmic number systems, as there exists no implicit dependency on a base-2 exponent for arithmetic operations. The sole immediate consequence of moving from base 2 to base $\sqrt{e}$ is a reduction of the dynamic range by a factor of $\log_2(\sqrt{e}) \approx 0.72$. It prompts scrutiny into how deeply ingrained the choice of base 2 is in logarithmic number systems, primarily due to the prevailing 'binarity' of computer architectures. In an alternate reality, where computers were constructed using ternary or alternative logic, discussions might revolve around the significance of base 3 or other bases for logarithmic number systems.

There has been recent pioneering work on exploring alternative bases for logarithmic number systems, particularly for tailored low precision applications, in [1]. However, it was only focused on tailoring dynamic range and expressive power, not arithmetic advantages. Takum represents the inaugural implementation of a pure base-$\sqrt{e}$ logarithmic number system with an explicit focus on improved arithmetic properties. The base $\sqrt{e}$ was chosen over e because the latter would yield an excessive dynamic range of $\pm(e^{-255}, e^{255}) \approx \pm(1.8 \times 10^{-111}, 5.6 \times 10^{110})$. Any arithmetic advantages provided by base e are however still accessible as a change from $\sqrt{e}$ to e and back merely constitutes a left and right shift of the fixed-point logarithmic value ℓ respectively.

Regardless of the underlying logic employed in a computing system, which is inherently of human design, the base e holds unparalleled significance due to its profound integration within mathematics and the natural sciences. Particularly, the pivotal role of the exponential function in mathematics, its utility in modelling growth, decay, and dynamic systems, alongside its widespread application across diverse fields including electromagnetism, quantum mechanics, thermodynamics, optics, acoustics, biological systems, and environmental science, among others, underscores the unique role of the base e in comparison to other bases. While e is an irrational number, it is worth noting that adopting a rational base such as 2 with non-integral exponents mostly yields irrational numbers as well. The principal advantage of employing a base of 2 lies in its capacity to accurately represent powers of two. Nonetheless, one must consider whether prioritising exact integral representations should be a fundamental design objective for a non-integral number system, or if it entails trade-offs. One might argue that an integer is effectively represented within a specific number system if a round-trip conversion-comprising conversion to this number system and back, followed by rounding to its original number of significant digits-results in the retrieval of the original integer.

The intricacies inherent to the GAUSSIAN logarithm remain ostensibly unaffected by alterations in the base when utilising lookup tables with interpolation, as such modifications merely involve a rescaling of the stored constants. However, it's noted that lookup tables with interpolation do not scale well to higher precisions beyond 32 bits [5]. The base e has been explored to pave the way for novel avenues in arithmetic [13], employing a mixed-base methodology (with a base 2 exponent and base e significand) alongside efficient $\ln()/\exp()$ evaluation algorithms sourced from [23]. These approaches are inapplicable to base 2 as the change of basis entails the multiplication or division of the result by $\ln(2)$, which is not power-efficient. In contrast, transitioning from the base e to the base $\sqrt{e}$ can be achieved with a simple bit shift of the stored logarithmic value, as $\ln(\sqrt{e}) = 0.5$. Given the constrained dynamic range characteristic of takums, adaptations are feasible for the entire logarithmic value ℓ, rather than confining oneself solely to a base $\sqrt{e}$ significand. The adaptation of the algorithms presented in [23] will be addressed in future works.

Another primary advantage of utilising a base-$\sqrt{e}$ logarithmic number system resides in the formulation of Gaussian logarithms (see (39)). Assuming that evaluating $\Phi_e^{\pm}$ is efficient, we observe that the expression

$$\Phi_{\sqrt{e}}^{\pm}(q) = \log_{\sqrt{e}}\left(1 \pm \sqrt{e}^{-q}\right) = 2\Phi_e^{\pm}\left(\frac{q}{2}\right) \tag{51}$$

involves minimal overhead, requiring just two bit shifts to transition between $\Phi_e^{\pm}$ and $\Phi_{\sqrt{e}}^{\pm}$. Conversely, in the conventional form with base 2, denoted as $\Phi_2^{\pm}$, we have

$$\Phi_2^{\pm}(q) = \log_2\left(1 \pm 2^{-q}\right) = \frac{\Phi_e^{\pm}(q\ln(2))}{\ln(2)}, \tag{52}$$

where rescaling with $\log(2)$ proves to be costly. This incurs overhead in the evaluation of hyperbolic and other elementary functions, often necessitating the application of ln(2) for argument or result rescaling – a practice commonly encountered due to the prevalent role of the natural base e in mathematics. Such considerations hold particular significance for evaluating activation functions in deep learning contexts [18], especially when combined with efficient evaluations of ln() and exp(), which warrant further exploration.

As a side note, it is worth highlighting the remarkable proximity between $\sqrt{e} \approx 1.65$ and the golden ratio $\frac{1+\sqrt{5}}{2} \approx 1.62$.

4.5   Formal Analysis

This work places particular emphasis on the formal verification of the proposed takum format, acknowledging the inherent superiority of formal analysis over manual mechanical verification. Formal verification not only facilitates the development of novel proof techniques, which were indispensable in substantiating certain propositions delineated within this section, but also contributes to maintaining rigor and precision in the analysis. In the interest of readability, most proofs are relegated to the appendix. The exposition commences with a straightforward proof, affirming the non-redundancy of number encodings in the takum format:

Proposition 3 (Takum Uniqueness). *Let $n \in \mathbb{N}_1$ and $B, \tilde{B} \in \{0,1\}^n$ as in Definition 2. It holds*

$$\tau(B) = \tau\left(\tilde{B}\right) \Rightarrow B = \tilde{B}, \tag{53}$$

which means that τ is an injective function.

Proof. See Section D.

Before proceeding to establish the next property, we first introduce some essential definitions:

Definition 3 (Unsigned Integer Mapping). *Let $n \in \mathbb{N}_1$ and an n-bit string $B = (B_{n-1}, \ldots, B_0) \in \{0,1\}^n$. The unsigned integer mapping $\mathrm{UI} \colon \{0,1\}^n \mapsto \{0, \ldots, 2^n - 1\}$ is defined as*

$$\mathrm{UI}(B) = \sum_{i=0}^{n-1} B_i 2^i. \tag{54}$$

Definition 4 (Bit String Incrementation). *Let $n \in \mathbb{N}_1$ and an n-bit string $B = (B_{n-1}, \ldots, B_0) \in \{0,1\}^n$. The bit string incrementation is defined as*

$$B + 1 := \begin{cases} \mathrm{UI}^{\mathrm{inv}}(\mathrm{UI}(B) + 1) & \mathrm{UI}(B) \in \{0, \ldots, 2^n - 2\} \\ 0 & \mathrm{UI}(B) = 2^n - 1. \end{cases} \tag{55}$$

Necessary for the subsequent proofs is the following elementary property of unsigned integers:

Lemma 1 (Unsigned Integer Negation). *Let $n \in \mathbb{N}_0$ and an n-bit string $B = (B_{n-1}, \ldots, B_0) \in \{0,1\}^n$. It holds*

$$\mathrm{UI}(\overline{B}) = 2^n - 1 - \mathrm{UI}(B). \tag{56}$$

Proof. By inserting Definition 3 and adding the right sum on both sides, we can see that (56) is equivalent to

$$2^n - 1 = \sum_{i=0}^{n-1} \left(\overline{B_i} + B_i\right) 2^i = \sum_{i=0}^{n-1} 2^i = 2^n - 1, \tag{57}$$

which proves it. □

After the introduction of unsigned integers, we proceed to define two's complement signed integers:

Definition 5 (Two's Complement Signed Integer Mapping). *Let $n \in \mathbb{N}_1$ and an n-bit string $B = (B_{n-1}, \ldots, B_0) \in \{0,1\}^n$. The two's complement signed integer mapping $\mathrm{SI}: \{0,1\}^n \mapsto \{-2^{n-1}, \ldots, 2^{n-1} - 1\}$ is defined as*

$$\mathrm{SI}(B) = -B_{n-1}2^{n-1} + \sum_{i=0}^{n-2} B_i 2^i. \tag{58}$$

Lemma 2 (Two's Complement Signed Integer Monotonicity). *Let $n \in \mathbb{N}_1$ and an n-bit string $B = (B_{n-1}, \ldots, B_0) \in \{0,1\}^n$ with $B \neq (0,1,\ldots,1)$. It holds*

$$\mathrm{SI}(B+1) = \mathrm{SI}(B) + 1. \tag{59}$$

Proof. For $(B_{n-2}, \ldots, B_0) \neq \mathbf{1}$, the bit string does not overflow upon incrementation. Similarly, the sum $\sum_{i=0}^{n-2} B_i 2^i$ does not overflow, thus validating the equation $\mathrm{SI}(B+1) = \mathrm{SI}(B) + 1$. The condition $B = (0,1,\ldots,1)$ has been explicitly excluded, leaving only the scenario where $B = (1,1,\ldots,1) = \mathbf{1}$. Consequently, it holds that

$$0 = \left(-2^{n-1} + 2^{n-1} - 1\right) + 1 = \mathrm{SI}(\mathbf{1}) + 1 = \mathrm{SI}(\mathbf{1} + 1) = \mathrm{SI}(\mathbf{0}) = 0, \tag{60}$$

as was to be shown. □

This monotonicity yields an order on $\{0,1\}^n$ induced by the image of the signed integer mapping SI, specifically the set $\left(\left\{-2^{n-1}, \ldots, 2^{n-1} - 1\right\}, \leq\right)$, which forms a subset of $(\mathbb{Z}, \leq)$, as follows:

Definition 6 (Two's Complement Signed Integer Partial Order). *Let $n \in \mathbb{N}_1$ and n-bit strings $B, \tilde{B} \in \{0,1\}^n$. The two's complement signed partial order is defined as*

$$B \preceq \tilde{B} \; :\longleftrightarrow \; \mathrm{SI}(B) \leq \mathrm{SI}\left(\tilde{B}\right), \tag{61}$$

yielding a partially ordered set $(\{0,1\}^n, \preceq)$.

With this apparatus established, we can formally articulate the concept of monotonicity concerning takums:

Proposition 4 (Takum Monotonicity). *Let $n \in \mathbb{N}_1$ and n-bit strings $B, \tilde{B} \in \{0,1\}^n \setminus \{(1,0,\ldots,0)\}$. It holds*

$$B \preceq \tilde{B} \longrightarrow \tau(B) \le \tau\left(\tilde{B}\right). \tag{62}$$

Proof. See Section E.

Having established the fundamental properties of takums, the focus now shifts towards demonstrating three essential bitwise operations: negation, inversion, and the combined operation of negation and inversion, applied to a given takum. Let us begin by examining the process of signed integer negation:

Proposition 5 (Two's Complement Signed Integer Negation). *Let $n \in \mathbb{N}_1$ and an n-bit string $B = (B_{n-1}, \ldots, B_0) \in \{0,1\}^n$ with $B \ne (1,0,\ldots,0)$. It holds*

$$\mathrm{SI}(\overline{B} + 1) = -\,\mathrm{SI}(B). \tag{63}$$

Proof. Given $B \ne (1,0,\ldots,0)$ we know that $\overline{B}+1$ never carries into the inverted MSB $\overline{B_{n-1}}$. For $B_{n-1} = 0$ it holds with Lemma 1

$$\mathrm{SI}(\overline{B} + 1) = -2^{n-1} + 2^{n-1} - 1 - \sum_{i=0}^{n-2} B_i 2^i + 1 =$$
$$-\left(0 \cdot 2^{n-1} + \sum_{i=0}^{n-2} B_i 2^i\right) = -\,\mathrm{SI}(B). \tag{64}$$

Likewise for $B_{n-1} = 1$ we can deduce

$$\mathrm{SI}(\overline{B} + 1) = -0 \cdot 2^{n-1} + 2^{n-1} - 1 - \sum_{i=0}^{n-2} B_i 2^i + 1 =$$
$$-\left(-1 \cdot 2^{n-1} + \sum_{i=0}^{n-2} B_i 2^i\right) = -\,\mathrm{SI}(B). \tag{65}$$

This was to be shown. □

Proposition 6 (Takum Negation). *Let $n \in \mathbb{N}_1$ and $(S, D, R, C, M) \in \{0,1\}^n$ as in Definition 2. It holds*

$$\tau\left((\overline{S}, \overline{D}, \overline{R}, \overline{C}, \overline{M}) + 1\right) = \begin{cases} -\tau((S, D, R, C, M)) & \tau((S, D, R, C, M)) \ne \mathrm{NaR} \\ \mathrm{NaR} & \tau((S, D, R, C, M)) = \mathrm{NaR}. \end{cases} \tag{66}$$

Proof. See Section F.

Most notably, this bitwise operation is tantamount to negating a two's complement integer. Subsequently, the ensuing observation serves as a minor yet pivotal simplification in elucidating the proof pertaining to the third bitwise operation under scrutiny.

Lemma 3 (Takum Inversion-Negation). *Let $n \in \mathbb{N}_1$ and $(S, D, R, C, M) \in \{0,1\}^n$ as in Definition 2 with $\tau((S, D, R, C, M)) \neq \text{NaR}$. It holds*

$$\tau\big((\overline{S}, D, R, C, M)\big) = \begin{cases} -\frac{1}{\tau((S,D,R,C,M))} & \tau((S, D, R, C, M)) \neq 0 \\ \text{NaR} & \tau((S, D, R, C, M)) = 0. \end{cases} \tag{67}$$

Proof. See Section G.

The final bitwise operation results in the inversion of a given takum. This aspect is particularly noteworthy when juxtaposed with posits, as the capability to perform this operation is unique to takums. This distinction arises from the intrinsic nature of takums as a logarithmic number system, wherein each numerical value possesses a perfect reciprocal.

Proposition 7 (Takum Inversion). *Let $n \in \mathbb{N}_1$ and $(S, D, R, C, M) \in \{0,1\}^n$ as in Definition 2 with $\tau((S, D, R, C, M)) \neq \text{NaR}$. It holds*

$$\tau\big((S, \overline{D}, \overline{R}, \overline{C}, \overline{M}) + 1\big) = \begin{cases} \frac{1}{\tau((S,D,R,C,M))} & \tau((S, D, R, C, M)) \neq 0 \\ \text{NaR} & \tau((S, D, R, C, M)) = 0. \end{cases} \tag{68}$$

Proof. See Section H.

For this property alone, it is prudent to embrace a logarithmic significand, thus achieving a symmetrical treatment of negation and inversion across the entirety of the numerical system.

Whilst the decoding process of a takum remains straightforward, encoding a floating-point value as a takum warrants further deliberation. We propose the following proposition for the lossless encoding and decoding of a given floating-point number:

Proposition 8 (Takum Floating-Point Encoding). *Let*

$$x := (-1)^s \cdot (1 + f) \cdot 2^h \in \left(-\sqrt{e}^{255}, -\sqrt{e}^{-255}\right) \cup \{0\} \cup \left(\sqrt{e}^{-255}, \sqrt{e}^{255}\right) \tag{69}$$

be a floating-point number with sign $s \in \{0,1\}$, fraction $f \in [0,1)$ and exponent $h \in \mathbb{Z}$ with

$$h \in \left(\frac{-127.5 - \ln(1+f)}{\ln(2)}, \frac{127.5 - \ln(1+f)}{\ln(2)}\right) \supset \{-184, 183\}. \tag{70}$$

Using the notation from Definition 2 it holds for $\tau((S, D, R, C, M)) = x$:

$$S = s, \tag{71}$$

$$\ell = 2\left(h\ln(2) + \ln(1 + f)\right) \in (-255, 255), \tag{72}$$

$$c = \left\lfloor (-1)^S \ell \right\rceil \in \{-255, \dots, 254\}, \tag{73}$$

$$D = c \geq 0, \tag{74}$$

$$r = \begin{cases} \lfloor \log_2(-c) \rfloor & D = 0 \\ \lfloor \log_2(c + 1) \rfloor & D = 1, \end{cases} \tag{75}$$

$$R = \begin{cases} 7 - r & D = 0 \\ r & D = 1, \end{cases} \tag{76}$$

$$C = \begin{cases} c + 2^{r+1} - 1 & D = 0 \\ c - 2^r + 1 & D = 1, \end{cases} \tag{77}$$

$$m = (-1)^S \ell - c \in [0, 1), \tag{78}$$

$$p = \inf_{i \in \mathbb{N}_0} \left(2^i m \in \mathbb{N}_0\right) \in \mathbb{N}_0 \cup \{\infty\}, \tag{79}$$

$$M = 2^p \, m \in \{0, 1\}^p. \tag{80}$$

We define $\tau^{\mathrm{inv}} \colon \{0, \mathrm{NaR}\} \cup \pm\left(\sqrt{e}^{-255}, \sqrt{e}^{255}\right) \mapsto \{0, 1\}^{5+r+p}$ *as*

$$\tau^{\mathrm{inv}}(x) := \begin{cases} (0, 0, \mathbf{0}, \mathbf{0}, \mathbf{0}) & x = 0 \\ (1, 0, \mathbf{0}, \mathbf{0}, \mathbf{0}) & x = \mathrm{NaR} \\ (S, D, R, C, M) & otherwise \end{cases} \tag{81}$$

Proof. See Section I.

Algorithms 2 and 3 delineate the encoding and decoding process for a non-zero floating-point number, adhering to the insights expounded in Proposition 8. It serves to illustrate that this procedure exhibits no greater complexity than that encountered in the encoding and decoding of (logarithmic) posits, which also require at least one evaluation of a logarithm and exponent respectively. Furthermore, there is no difference in computational complexity between using base 2 or $\sqrt{e}$.

Algorithm 2: Takum encoding algorithm for a floating-point number of the form $(-1)^s \cdot (1+g) \cdot 2^h \in \pm \left(\sqrt{e}^{-255}, \sqrt{e}^{255} \right)$ based on Proposition 8.

---

input : $s \in \{0,1\}$: sign
       $f \in [0,1)$: fraction
       $h \in \mathbb{Z}$ as in (70): exponent
output: S : sign bit
       D : direction bit
       R : regime bits
       r : regime value
       C : characteristic bits
       $p \in \mathbb{N}_0 \cup \{\infty\}$:
         mantissa bit count
       $M := (M_{p-1}, \ldots, M_0)$:
         mantissa bits

$S \leftarrow s$
$\ell \leftarrow 2\,(h \ln(2) + \ln(1+f))$

$c \leftarrow \left\lfloor (-1)^S \ell \right\rceil$
if $c \geq 0$ **then**
    $D \leftarrow 1$
    $r \leftarrow \lfloor \log_2(c+1) \rfloor$
    $R \leftarrow r$
    $C \leftarrow a - 2^r + 1$
else
    $D \leftarrow 0$
    $r \leftarrow \lfloor \log_2(-c) \rfloor$
    $R \leftarrow 7 - r$
    $C \leftarrow a + 2^{r+1} - 1$
end
$m \leftarrow (-1)^S \ell - c$
$p \leftarrow \inf_{i \in \mathbb{N}_0} \left(2^i m \in \mathbb{N}_0 \right)$
$M \leftarrow 2^p m \in \{0,1\}^p$

Algorithm 3: Floating-point encoding algorithm of a takum of the form $(-1)^S \sqrt{e}^\ell \in \pm \left(\sqrt{e}^{-255}, \sqrt{e}^{255} \right)$ to $(-1)^s \cdot (1+g) \cdot 2^h \in \pm \left(\sqrt{e}^{-255}, \sqrt{e}^{255} \right)$.

---

input : $S \in \{0,1\}$: sign
       $\ell \in (-255, 255)$:
         logarithmic value
output: $s \in \{0,1\}$: sign
       $f \in [0,1)$: fraction
       $h \in \mathbb{Z}$: exponent

$s \leftarrow S$
$h \leftarrow \lfloor \ell \rfloor$
$f \leftarrow \exp\!\left(\tfrac{1}{2}(\ell - h) \right) - 1$

Another crucial aspect for formal analysis is error analysis. A key quantity in numerical analysis is the constant upper bound,

$$\left| \frac{x - \mathrm{fl}(x)}{x} \right| \leq 2^{-n_f - 1} := \varepsilon(n_f), \tag{82}$$

of the relative approximation error of a number $x \in \mathbb{R}$ within the normal range of an IEEE 754 floating-point format with n_f fraction bits (refer to Table 1). Here, fl denotes the rounding operation. A similar bound applies to posits with an upper limit of $\varepsilon(p)$, where $p \in \{0, \ldots, n-5\}$ (as seen in (8)) represents the number of fraction bits, which varies due to the tapered exponent. Since p is unbounded from below, the relative approximation error can potentially be up to 50% irrespective of n. This characteristic poses challenges in applying standard numerical analysis techniques to posits in general. Furthermore, there lacks a theoretical framework to comprehend the comparatively higher precision of tapered floating-point formats exhibited by numbers near unity, confining posits to empirical performance enhancements without formal guarantees. In contrast, takums enforce a lower limit on the number of available mantissa bits, possibly enabling the application of standard numerical analysis techniques until a theoretical framework for tapered floating-point arithmetic is developed. This will be investigated as follows.

For $n \geq 12$ and a given $X \in \{0,1\}^n$ with $x = \tau(X) \notin \{0, \mathrm{NaR}\}$, it is possible to determine the precise count of available mantissa bits p (refer to (20)) solely based on the represented takum value:

Proposition 9 (Takum Mantissa Bit Count). *Let $n \in \mathbb{N}$ with $n \geq 12$ and $X \in \{0,1\}^n$ with $x := \tau(X) \notin \{0, \mathrm{NaR}\}$ and mantissa bit count p as in Definition 2. It holds*

$$p = n - 5 - \lfloor \log_2(|\lfloor 2\ln(|x|)\,\mathrm{sign}(x)\rfloor + (\ln(|x|)\,\mathrm{sign}(x) \geq 0)|)\rfloor \in$$
$$\{n - 12, \ldots, n - 5\}. \quad (83)$$

Proof. See Section J.

While this outcome may initially appear to offer limited utility, it can be extended to encompass all numbers within the dynamic range, including those that cannot be directly represented. Such an extension permits the determination of the guaranteed number of mantissa bits for any given number prior to the commencement of rounding operations.

Proposition 10 (Takum Mantissa Bit Count Lower Bound). *Let $x \in \pm\left(\sqrt{e}^{-255}, \sqrt{e}^{255}\right)$ and $n \in \mathbb{N}$ with $n \geq 12$. It holds for $X \in \{0,1\}^n$ with $\mathrm{round}_n(x) = \tau(X)$ and mantissa bit count p as in Definition 2*

$$p \geq n - 6 - \lfloor \log_2(|\lfloor 2\ln(|x|)\,\mathrm{sign}(x)\rfloor + (\ln(|x|)\,\mathrm{sign}(x) \geq 0)|)\rfloor \in$$
$$\{n - 13, \ldots, n - 6\}. \quad (84)$$

Proof. See Section K.

This finding holds significant utility as it furnishes us with the capability to gauge the precision of any given number when expressed as a takum even prior to rounding. Such an outcome assumes particular importance, unlike in the realm of uniform precision arithmetic such as IEEE 754 floating-point numbers, where the relative precision remains consistent across all normal numbers. Upon scrutinising the proof, it becomes evident that alterations in the mantissa bit count occur only exceptionally rarely during rounding processes. It may be feasible in subsequent research endeavours to establish an even more robust lower bound for the mantissa bit count.

Having established both the mantissa bit count and its lower bound for arbitrary numbers, we are now poised to delve into an examination of the relative approximation error inherent in takum arithmetic. Specifically, we shall derive an upper bound for the relative approximation error contingent upon the mantissa bit count p:

Proposition 11 (Takum Machine Precision). *Let* $x \in \pm \left(\sqrt{e}^{-255}, \sqrt{e}^{255} \right)$, $n \geq 12$ *and* $X \in \{0,1\}^n$ *with* $\text{round}_n(x) = \tau(X)$. *The bit string* X *has the mantissa bit count* $p \in \{n - 12, \ldots, n - 5\}$ *as in Definition 2. It holds for the relative approximation error*

$$\left| \frac{x - \text{round}_n(x)}{x} \right| \leq \sqrt{e}^{2^{-p-1}} - 1 =: \lambda(p). \tag{85}$$

The upper bound $\lambda(p)$ *satisfies*

$$\lambda(p) < \frac{2}{3}\varepsilon(p) < \varepsilon(p). \tag{86}$$

Proof. See Section L.

Because takums constitute a logarithmic number system, the upper bound assumes a distinct format compared to IEEE 754 floating-point numbers and posits. However, the most noteworthy disparity, particularly when contrasted with posits, lies in the fact that the mantissa bit count, denoted as p, is constrained to a minimum of $n - 12$. Consequently, the relative approximation error of takums is bounded above by $\lambda(n - 12) < \frac{2}{3}\varepsilon(n - 12)$ for $n \geq 12$, a value readily applicable in standard numerical analysis theory. Despite its modest quality, this upper bound establishes a nexus with the extensive body of literature surrounding IEEE 754 floating-point numbers. Moreover, in the context of `float64`, which boasts $n_f = 52$ fraction bits, `takum64` assures the same minimal mantissa bit count of $n - 12 = 52$. Given $\lambda(52) < \frac{2}{3}\varepsilon(52)$ it can be deduced that `takum64` presents only at most two-thirds of the relative approximation error of `float64`, within the dynamic range of `takum64`. This observation in turn means that `takum64` can be presumed to possess at least the same (uniform) machine precision to `float64`, thereby enabling the direct application of all results pertinent to double-precision IEEE 754 floating-point numbers that depend upon machine precision.

Future research endeavours shall explore novel methodologies for analysing the tapered precision inherent in posits and takums, aiming to comprehensively elucidate and formalise the advantages associated with tapered precision numerical formats. This nascent area of inquiry may be designated as 'tapered precision numerical analysis', diverging from the prevailing paradigm of 'uniform precision numerical analysis'. Nonetheless, until such investigations yield substantive results, takums offer a distinct advantage over posits by enabling the application of theoretical frameworks predicated on the assumption of a constant relative approximation error.

4.6   NaR Convention

Conveniently omitted in Proposition 4 is the role of NaR in the ordering of takums. This is due to the discretionary nature of NaR handling. The IEEE 754 standard encompasses various forms of NaN and ultimately elected to stipulate NaN $\neq$ NaN universally, primarily because its initial specification did not mandate implementations to furnish a mechanism for discerning if a given floating-point number is a NaN. Instead, users were offered a recourse through the distinctive property $x \neq x \rightarrow x = \text{NaN}$ to identify NaNs [15, 8]. Analogously, all comparisons involving NaNs yield `false`. Blindly espousing this convention for takums sans introspection would be imprudent.

In the 2019 revision of the IEEE 754 standard, a total-ordering predicate was introduced (see [12, §5.10]), which subtly alters the treatment of NaN. Notably, in addition to other modifications irrelevant to takums, the equality NaN = NaN is upheld, while $-$NaN is deemed smaller than the smallest representable number, and NaN is considered larger than the largest representable number. These adjustments suggest that the original handling of NaN may not be optimal. Consequently, we propose the following convention for managing NaRs within the context of takums:

Definition 7 (NaR **Total-Ordering Convention**). *Let* $n \in \mathbb{N}_1$. *NaR is defined according to the total-ordering convention if and only if*

$$(\text{NaR} = \text{NaR}) \wedge \left(\forall_{\text{NaR} \neq x \in \tau(\{0,1\}^n)} \colon \text{NaR} < x \right) \tag{87}$$

hold.

While there is no universally optimal method for defining NaR handling, establishing it in a manner that ensures the takum number system maintains a total order is deemed reasonable. Such an approach guarantees that NaR = NaR remains valid and that NaR is deemed smaller than the smallest representable number. This alignment is in harmony with the convention regarding NaR in posits [11] and the total-ordering predicate in IEEE 754-2019 [12, §5.10].

This NaR convention proves to be judicious for hardware implementations, owing to the fact that the bit representation of NaR, $(1, 0, \ldots, 0)$, coincides with that of the smallest two's complement signed integer. Consequently, no special case for comparisons is necessitated, and takums can be compared akin to two's complement signed integers. It is also a pragmatic decision to designate NaR as the smallest representable number for practical applications. This choice finds resonance in commonplace approximation loops, typified by constructs such as `while (residual < bound)`, where a predetermined bound triggers termination upon encountering NaR as the residual value, which is desirable.

4.7 Linear Takums

While takums are categorised within the ambit of logarithmic number systems, boasting numerous advantageous properties and the innovative application of a base $\sqrt{e}$ approach which heralds the promise of more efficient hardware implementations for applications demanding higher precision, it is pertinent to acknowledge that the realm of logarithmic number systems remains comparatively nascent. This domain embodies a paradigm-shifting perspective that necessitates a fundamental reevaluation of the conventional tenets underpinning floating-point arithmetic. Amidst this discourse, the utility of logarithmic significands in comparison to their linear counterparts remains an open question, poised at the brink of being either a revolutionary advancement or a notable misstep in the quest for optimizing arithmetic computation.

Given that the logarithmic significand is not the sole distinguishing aspect of takums, and considering that the primary feature, namely the efficient characteristic encoding, likely holds benefits even in a traditional floating-point context, we will now define takums with a linear base-2 significand for applications that favour a floating-point representation. As detailed in Sect. 4.3, the terms 'characteristic' and 'mantissa' are deemed inappropriate for non-logarithmic number systems. Consequently, the bit string segments and variables are renamed accordingly to refer to 'exponent' and 'fraction', respectively. However, the definition of the characteristic is retained as it facilitates the articulation of an exponent that precisely aligns with the exponent of the floating-point representation.

Definition 8 (linear takum encoding). *Let $n \in \mathbb{N}$ with $n \geq 12$. Any n-bit MSB→LSB string $(S, D, R, C, F) \in \{0,1\}^n$ of the form*

with

$$S \in \{0,1\} \qquad\qquad : sign\ bit \qquad (88)$$

$$D \in \{0,1\} \qquad\qquad : direction\ bit \qquad (89)$$

$$R := (R_2, R_1, R_0) \in \{0,1\}^3 \qquad : regime\ bits \qquad (90)$$

$$r := \begin{cases} 7 - (4R_2 + 2R_1 + R_0) & D = 0 \\ 4R_2 + 2R_1 + R_0 & D = 1 \end{cases} \in \{0,\dots,7\} \quad : regime \qquad (91)$$

$$C := (C_{r-1}, \dots, C_0) \in \{0,1\}^r \qquad : characteristic\ bits \quad (92)$$

$$c := \begin{cases} -2^{r+1} + 1 + \sum_{i=0}^{r-1} C_i 2^i & D = 0 \\ 2^r - 1 + \sum_{i=0}^{r-1} C_i 2^i & D = 1 \end{cases} \qquad : characteristic \qquad (93)$$

$$p := n - r - 5 \in \{n - 12, \dots, n - 5\} \qquad : fraction\ bit\ count \quad (94)$$

$$F := (F_{p-1}, \dots, F_0) \in \{0,1\}^p \qquad : fraction\ bits \qquad (95)$$

$$f := 2^{-p} \sum_{i=0}^{p-1} F_i 2^i \in [0,1) \qquad\qquad : fraction \qquad (96)$$

$$e := (-1)^S (c + S) \in \{-255, \dots, 254\} \qquad : exponent \qquad (97)$$

encodes the linear takum value

$$\overline{\tau}((S, D, R, C, F)) := \begin{cases} \begin{cases} 0 & S = 0 \\ \text{NaR} & S = 1 \end{cases} & D = R = C = F = 0 \\ [(1 - 3S) + f] \cdot 2^e & otherwise \end{cases} \qquad (98)$$

with $\overline{\tau} \colon \{0,1\}^n \mapsto \{0, \mathrm{NaR}\} \cup \pm \left(2^{-255}, 2^{255}\right)$. *Without loss of generality, any bit string shorter than 12 bits is also considered in the definition by assuming the missing bits to be zero bits ('ghost bits').*

Implementers are at liberty to adopt either variant for takums, albeit the logarithmic significand in Definition 2 is designated as the standard. It is incumbent upon implementations to explicitly specify whether they support 'linear takums' or 'logarithmic takums'. In the absence of such clarification, the logarithmic variant is to be presumed.

The linear takums more closely adhere to the original definition of posits outlined in Definition 1 than the logarithmic takums. However, the posit definition exhibits a minor deficiency in failing to define a variable for the actual floating-point exponent. To address this, we introduce the term 'characteristic' and designate variable c to represent the 'pre-exponent', while defining the exponent e to denote the actual floating-point exponent.

While this document delineates the concept of linear takums, they will not be encompassed within the ensuing evaluation. Nonetheless, it is imperative to acknowledge that the cornerstone findings of the formal analysis – specifically Propositions 3 (uniqueness), 4 (monotonicity, thus also the NaR convention elaborated in Sect. 4.6), and 6 (negation) – remain applicable regardless of the significand being linear or logarithmic. Linear takums have the relative approximation error $\varepsilon(p)$ and we can define a linear takum rounding function $\overline{\mathrm{round}}_n(x)$ as in Algorithm 1 by adapting the bounds to base 2. Algorithms 4 and 5 delineate the procedures for converting a floating-point number to a linear takum and vice versa, respectively. While these algorithms are presented without formal proof, their derivation follows directly from the definitions and results from equating $(-1)^s(1+g)2^h$ and $[(1-3S)+f] \cdot 2^e$. This equivalence necessitates the borrowing or lending of one factor of 2 in cases where g or f equals zero.

Upon examination, it becomes apparent that linear takums do not significantly lag behind logarithmic takums in terms of analytical outcomes. In the event that logarithmic number systems fail to gain traction, linear takums present themselves as a feasible alternative. However, there exist two distinct aspects where linear takums exhibit shortcomings: firstly, they lack a straightforward bitwise inversion mechanism, as demonstrated in Proposition 7; secondly, their machine precision is at least two-thirds inferior to that of logarithmic takums, as indicated in Proposition 11. Notably, the former deficiency, which implies a mathematical symmetry between negation and inversion, holds particular appeal for its elegance.

5   Evaluation

In the ensuing analysis, we will appraise takums across various dimensions to assess their efficacy both as a numerical system for general-purpose arithmetic, juxtaposed against IEEE 754 floating-point numbers, and as a tapered floating-point format, juxtaposed against posits.

5.1 GUSTAFSON Criteria

We commence by examining our adherence to the GUSTAFSON criteria as delineated in Sect. 1.1. Given our deliberate design of takums to conform to the dynamic range criteria, outlined in Sect. 1.2, there is no necessity to scrutinise them further.

We fulfil *Property 1 (distribution)* by establishing a rational dynamic range of numbers, detailed in Sect. 1.2, while the tapered format of takums aptly encompasses a greater proportion of numbers proximal to 1, aligning with typical computational requirements. *Property 2 (uniqueness)* has been substantiated, as evidenced by Proposition 3. *Property 3 (generality)* is inherent in our construction, with no imposed restrictions on the bit string length n. By construction the satisfaction of *Property 4 (statelessness)* is also assured. *Property 5 (exactness)*, though intricately linked with implementation specifics, finds support in the encouraging outcomes detailed in Sect. 5.6.

Algorithm 4: Linear takum encoding algorithm of a floating-point number of the form $(-1)^s \cdot (1 + g) \cdot 2^h \in \pm \left(2^{-255}, 2^{255}\right)$.

input : $s \in \{0, 1\}$: sign
$g \in [0, 1)$: fraction
$h \in \{-255, 254\}$: exponent
$h = -255 \rightarrow g \neq 0$

output: S : sign bit
D : direction bit
R : regime bits
C : amplitude bits
$F := (F_{m-1}, \ldots, F_0)$:
fraction bits

$S \leftarrow s$
if $S = 0$ **then**
$\quad c \leftarrow h$
$\quad f \leftarrow g$
else
$\quad$ **if** $g = 0$ **then**
$\quad\quad c \leftarrow -h$
$\quad\quad f \leftarrow 0$
$\quad$ **else**
$\quad\quad c \leftarrow -h - 1$
$\quad\quad f \leftarrow 1 - g$
$\quad$ **end**
end

$D \leftarrow c \geq 0$
if $D = 0$ **then**
$\quad r \leftarrow \lfloor \log_2(-c) \rfloor$
$\quad R \leftarrow 7 - r$
$\quad C \leftarrow c + 2^{r+1} - 1$
else
$\quad r \leftarrow \lfloor \log_2(c + 1) \rfloor$
$\quad R \leftarrow r$
$\quad C \leftarrow c - 2^r + 1$
end
$p \leftarrow \inf_{i \in \mathbb{N}_0} \left(2^i f \in \mathbb{N}_0\right)$
$F \leftarrow 2^p f$

Algorithm 5: Floating-point encoding algorithm of a linear takum of the form $[(1-3S)+f]\cdot2^e \in \pm(2^{-255}, 2^{255})$ to $(-1)^s\cdot(1+g)\cdot2^h \in \pm\left(2^{-255}, 2^{255}\right)$.

input : $S \in \{0,1\}$: sign
        $f \in [0,1)$: fraction
        $e \in \{-255, 254\}$: exponent
        $e = -255 \rightarrow f \neq 0$
output: $s \in \{0,1\}$: sign
        $g \in [0,1)$: fraction
        $h \in \{-255, 254\}$: exponent
        $h = -255 \rightarrow g \neq 0$

$s \leftarrow S$
if $s = 0$ then
   | $h \leftarrow e$
   | $g \leftarrow f$
else
   if $f = 0$ then
      | $h \leftarrow e + 1$
      | $g \leftarrow 0$
   else
      | $h \leftarrow e$
      | $g \leftarrow 1 - f$
   end
end

Property 6 (binary monotonicity) is formally established in Proposition 4, mirroring the proof for *Property 7 (binary negation)* as elucidated in Proposition 6. Additionally, we affirm the property of binary inversion via Proposition 7. Regarding *Property 8 (flexibility)*, we contend that our approach potentially exceeds posits' capabilities, as not only can bit strings of varying lengths be effortlessly converted to other lengths, but uniform decoding logic can be applied across all variants. *Property 9 (NaR propagation)* pertains to implementation intricacies, while the discussion surrounding *Property 10 (implementation simplicity)* is reserved for Sect. 5.2, albeit preliminary satisfaction is inferred. In sum, our adherence to the GUSTAFSON criteria is comprehensive.

5.2   Hardware Implementation

Despite the ostensibly more intricate mathematical definition of takums compared to posits (refer to Definitions 2 and 1), the former's hardware implementation is considerably simpler. With merely 8 regime states, a lookup-table necessitates only 3 entries per state for comprehensive format parsing: a bit mask for direct characteristic bits extraction (12 bits per entry, with the implicit 1 specified in the mask and a consistent 5-bit left shift per entry) and the offset $5 + r$ denoting the initiation of fraction bits (4 bits per entry), if not directly computed. This culminates in a modest LUT size of 16 bytes. Given the characteristic bits' confinement within the initial 12 bits, logic application is also only necessary in this restricted domain. Moreover, the identical exponent parsing logic and lookup-tables can be universally employed across all types, facilitating encoding processes (refer to Algorithm 2), which have been demonstrated to be straightforward, no more intricate than posit encoding.

In contrast, the exponent in posit format may span the entire posit width, necessitating logic attachment to all input bits for exponent parsing. Determining k mandates a resource-intensive bit-counting methodology, thereby also impeding posit software implementations. If regime detection were to be executed through lookup-tables, the table size would grow with increasing bit string lengths.

Logarithmic fractions and their hardware complexity have been extensively studied [2,5,6]. Recent advancements and hardware implementations demonstrate that addition and subtraction can be performed with comparable or even lower latency than floating-point operations using linear significands. For all other arithmetic operations, logarithmic significands exhibit significantly reduced latency and power consumption [5,13]. From a practical standpoint, when implementing an Arithmetic Processing Unit (APU) for handling logarithmic fractions, focus solely on addition and subtraction is necessary, eliminating the need for implementing and optimizing complex operations such as multiplication, division, square root extraction, and squaring, as these operations can be reduced to fixed-point additions, subtractions, and shifts. This aspect is further underscored by formally proven bitwise operations enabling rapid negation and inversion of a takum, thus obviating the necessity for dedicated inverter logic. For scenarios requiring higher bit counts where LUT based approaches become impractical, existing methodologies, such as those discussed in [13], can be adapted. This adaptation leverages the novel foundational principle that takums are based on the basis $\sqrt{e}$.

Additionally, by adhering to the NaR convention as elucidated in Sect. 4.6, the dedicated handling of NaR values becomes superfluous without compromising mathematical integrity. Instead, takums can be seamlessly type-cast to two's complement signed integers for comparison across all bit representations, including instances featuring NaR.

5.3   Coding Efficiency

In considering coding efficiency, our investigation focuses on the bit count necessary to encode a given positive number, without loss of generality. In the case of takums, this pertains to encoding the characteristic c, the integral component of the logarithmic value ℓ. A thorough comparative analysis of takums vis-à-vis other encoding schemes is presented in Fig. 3. The efficacy of coding is evaluated by examining the intrinsic encoding capabilities of the schemes when applied to numerals. This methodology deliberately avoids direct comparisons based on exponents, as such an approach would disproportionately benefit formats other than takums. This discrepancy arises because exponentiation to the base $\sqrt{e}$ increases at a more gradual pace compared to exponentiation to the base 2. When interpreting the results, it is imperative to discern between the performance concerning small and large values. Depending on the application, the significance of one over the other varies. In the realm of small values $(0 - 15)$, posits maintain an overall superiority, surpassing all alternative methods. Particularly, the ELIAS codes exhibit considerable initial overhead. Takums rank second, closely matching the performance of posits initially, albeit with a slight degradation of around 1 bit overall, ultimately outperforming ELIAS codes beyond the number 2.

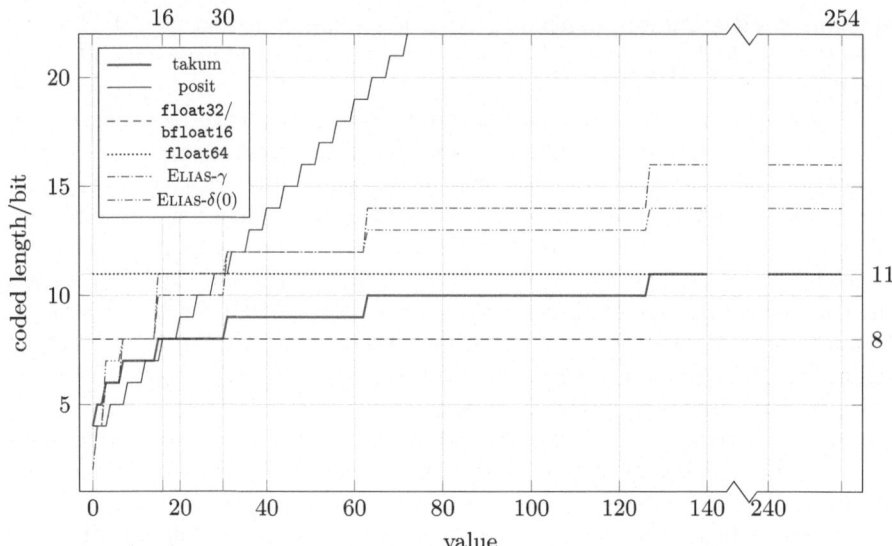

Fig. 3. Bit requirements for encoding a given value with different methods. The ELIAS codes are both defined in [20].

Conversely, in the context of higher values (> 15), takums demonstrate superior coding efficiency, notably 'overtaking' posits at number 16 with an additional 8 bits, comparable to the `bfloat16`/`float32` encoding cost. Takum encoding remains at 8 bits until number 30 (where posits already require 11 bits), never exceeding 11 bits (comparable to `float64`) until the highest value 254, where it also demonstrates to surpass the ELIAS codes, which necessitate 14 and 16 bits respectively. Posits, as previously noted, exhibit subpar performance for large numbers, a contributing factor to their unsuitability for general-purpose arithmetic.

The ELIAS codes, previously identified as possible alternatives to the posit encoding scheme [20], scale much better, but only manage to 'overtake' posits at numbers 31 with 12 bits, which is 1 bit more than `float64`, and are significantly less efficient than the takum encoding overall. Thus even though the ELIAS codes exhibit better asymptotic behaviour than posits, their initial and general prefix-code overhead is too high. Overall when compared to IEEE 754 floating-point numbers, it is evident that takums are at least as efficient as `float32`/`bfloat16` for numbers up to 30 and outperform or at least match `float64` up to the maximum value 254. It shall be noted here that `float32`/`bfloat16`, based on the previous discussion in this paper, do not offer a suitable dynamic range for general purpose arithmetic.

These findings underscore that takums strike a fine balance between small and large values, rendering them apt for general-purpose arithmetic, while also incorporating the advantages of tapered floating-point arithmetic previously associated with posits [10].

5.4   Dynamic Range

As illustrated in Fig. 4, takums exhibit a consistent dynamic range across various bit-lengths, efficiently achieving this range even with a limited number of bits. In contrast, posits only achieve a comparable dynamic range with more than 47 bits, while IEEE 754 floating-point numbers demonstrate inadequate dynamic range for 32 bits and fewer, and excessive dynamic range for 64 bits and beyond. Notably, the proprietary formats `bfloat16` and `TF32` display insufficient dynamic range to serve as general-purpose arithmetic formats.

Figure 4 further demonstrates that subnormal numbers insignificantly extend the dynamic range of IEEE 754 floating-point numbers, which elucidates why `bfloat16` and `TF32` have omitted them to reduce overhead and questions their overall utility.

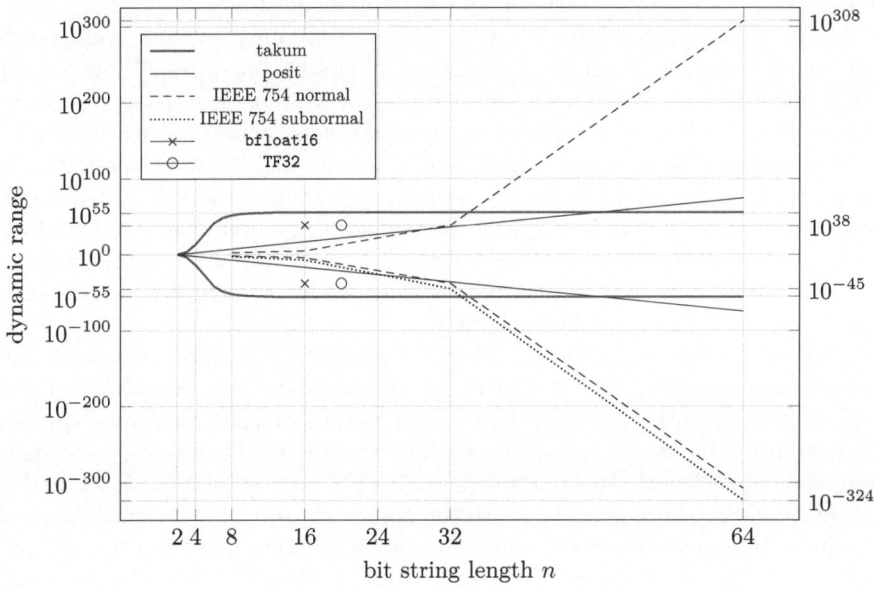

Fig. 4. Dynamic range comparison between various number formats relative to the bit string length n.

5.5   Absolute and Relative Approximation Error

In this section, we explore both the absolute and relative approximation errors of takums compared to other numerical formats. Drawing insights from the discussion in Sect. 1.2 and inspired by [7], we conduct benchmark tests on takums, posits, and IEEE 754 floating-point formats across various bit lengths using a set of physical constants. The first group comprises the six constants defining the

International System of Units (SI), namely the PLANCK constant h, the BOLTZ-MANN constant k, the elementary charge e (not to be confused with EULER's number), the speed of light c, the caesium standard $\Delta \nu$, and the AVOGADRO constant N_A (refer to Table 5). The second group encompasses two constants of vastly differing magnitudes: the cosmological constant Λ and the mass of the universe M (refer to Table 6). All values are rounded to the same number of significant digits as the ground truth values and are presented without units for conciseness.

It should be noted that these benchmarks are not intended to be the forte of tapered formats. On the contrary, [7] introduced this approach to highlight the shortcomings of posits when dealing with numbers of extremely small or large magnitudes. This is where uniform precision formats, such as IEEE 754, hold an advantage. Therefore, the aspiration with tapered precision formats like posits and takums is, at the very least, to achieve performance comparable to that of uniform precision formats with an equivalent number of bits.

In the first group (Table 5), it is evident that only `takum8` consistently matches the magnitudes of the constants among all 8-bit types. While `float8` either underflows or overflows, `posit8`'s employment of saturation arithmetic unsuccessfully strives to provide an answer that is as accurate as feasible. With 16 bits, `takum16` demonstrates comparable to but slightly worse performance than `bfloat16`, despite the former possessing a roughly 50% larger dynamic range. Conversely, `posit16` displays inferior accuracy overall and suffers from inadequate dynamic range for most constants. Across 19 and 32 bits, takums and IEEE 754 floating-point numbers are on par in performance, while posits consistently exhibit diminished accuracy.

The second group (Table 6) further emphasises the dynamic range of each number format. Among the 8-bit formats, none can accurately express the magnitude of the mass of the universe, though `takum8` comes closest and even expresses the magnitude of the cosmological constant accurately. The disparity becomes more pronounced with 16 bits: whereas `float16` and `bfloat16` either underflow or overflow to 0 and ∞ respectively, and `posit16` yields its minimum and maximum representable values, which deviate approximately 40 orders of magnitude from each constant, `takum16` successfully captures the magnitude of both constants and even two significant digits. This trend persists with 19 bits.

At 32 bits, `float32` still fails to represent either constant accurately, while `posit32` remains around 20 orders of magnitude adrift. In stark contrast, `takum32` precisely represents both constants with all significant digits intact, particularly noteworthy as the cosmological constant involves five significant digits. This exemplifies the inadequacy of `bfloat16`, TF32, `float32`, and `posit32` in handling general-purpose arithmetic, especially given the likelihood of encountering such large numbers as intermediate computational results (e.g., when squaring the Boltzmann constant), underscoring the potential of `takum32` and takums in general as general purpose arithmetic formats.

Regarding the relative approximation error, an upper bound λ for the relative approximation error of takum has been derived in Proposition 11, which

Table 5. Comparison of representations of the SI-defining constants in various formats.

name	PLANCK constant	BOLTZMANN constant	elementary charge
symbol	h	k	e
value	$6.62607015 \times 10^{-34}$	1.380649×10^{-23}	$1.602176634 \times 10^{-19}$
float8	0	0	0
posit8	$5.96046448 \times 10^{-8}$	5.960464×10^{-8}	$5.960464478 \times 10^{-8}$
takum8	$2.97569687 \times 10^{-35}$	4.303623×10^{-23}	$1.282891824 \times 10^{-19}$
float16	0	0	0
bfloat16	$6.62038418 \times 10^{-34}$	1.385528×10^{-23}	$1.600892270 \times 10^{-19}$
posit16	$1.38777878 \times 10^{-17}$	1.387779×10^{-17}	$1.387778781 \times 10^{-17}$
takum16	$6.56428218 \times 10^{-34}$	1.375520×10^{-23}	$1.596584671 \times 10^{-19}$
TF32	$6.62790735 \times 10^{-34}$	1.380358×10^{-23}	$1.601951062 \times 10^{-19}$
posit19	$3.38813179 \times 10^{-21}$	3.388132×10^{-21}	$2.168404345 \times 10^{-19}$
takum19	$6.61576649 \times 10^{-34}$	1.380904×10^{-23}	$1.602833526 \times 10^{-19}$
float32	$6.62607018 \times 10^{-34}$	1.380649×10^{-23}	$1.602176598 \times 10^{-19}$
posit32	$7.70371978 \times 10^{-34}$	1.380358×10^{-23}	$1.602215759 \times 10^{-19}$
takum32	$6.62607126 \times 10^{-34}$	1.380649×10^{-23}	$1.602176753 \times 10^{-19}$
name	speed of light	caesium standard	AVOGADRO constant
symbol	c	$\Delta\nu$	N_A
value	2.99792458×10^{8}	$9.192631770 \times 10^{9}$	$6.02214076 \times 10^{23}$
float8	∞	∞	∞
posit8	1.67772160×10^{7}	$1.677721600 \times 10^{7}$	1.67772160×10^{7}
takum8	2.94267566×10^{8}	$1.606646472 \times 10^{10}$	$1.26865561 \times 10^{24}$
float16	∞	∞	∞
bfloat16	2.99892736×10^{8}	$9.193914368 \times 10^{9}$	$6.02101727 \times 10^{23}$
posit16	3.01989888×10^{8}	$9.663676416 \times 10^{9}$	$7.20575940 \times 10^{16}$
takum16	2.98901606×10^{8}	$9.226194467 \times 10^{9}$	$5.99270479 \times 10^{23}$
TF32	2.99892736×10^{8}	$9.193914368 \times 10^{9}$	$6.02101727 \times 10^{23}$
posit19	2.99892736×10^{8}	$9.126805504 \times 10^{9}$	$2.95147905 \times 10^{20}$
takum19	2.99778578×10^{8}	$9.190224944 \times 10^{9}$	$6.02792137 \times 10^{23}$
float32	2.99792448×10^{8}	$9.192631296 \times 10^{9}$	$6.02214064 \times 10^{23}$
posit32	2.99792384×10^{8}	$9.192636416 \times 10^{9}$	$6.02101727 \times 10^{23}$
takum32	2.99792444×10^{8}	$9.192632204 \times 10^{9}$	$6.02214098 \times 10^{23}$

varies depending on the number of available fraction bits. In contrast, IEEE 754 floating-point numbers and posits, both featuring linear significands rather than logarithmic ones like takum, are characterised by the well-known ε (refer to (82))

Table 6. Comparison of representations of two very large physical constants in various formats.

name	cosmological constant	mass of the universe
symbol	Λ	M
value	1.1056×10^{-52}	1.5×10^{53}
float8	0	∞
posit8	5.9605×10^{-8}	1.7×10^{7}
takum8	1.2642×10^{-52}	7.9×10^{51}
float16	0	∞
bfloat16	0	∞
posit16	1.3878×10^{-17}	7.2×10^{16}
takum16	1.1156×10^{-52}	1.5×10^{53}
TF32	0	∞
posit19	3.3881×10^{-21}	3.0×10^{20}
takum19	1.1070×10^{-52}	1.5×10^{53}
float32	0	∞
posit32	7.5232×10^{-37}	1.3×10^{36}
takum32	1.1056×10^{-52}	1.5×10^{53}

as an upper bound for the relative approximation error. This bound remains constant for each type of IEEE 754 floating-point number, while it varies for posits based on the number of available fraction bits. Proposition 11 has demonstrated that, for the same number of available fraction bits, λ is smaller than $\frac{2}{3}\varepsilon$, indicating that logarithmic significands offer better relative accuracy.

Subsequently, we proceed to explore the upper bounds λ and ε for different bit lengths. In the ensuing plots, we depict the negative binary logarithm of λ and ε. Consequently, the graphs are inverted, with higher values denoting greater precision. Starting with 8 bits, we conduct a comparative analysis among takum8, posit8, and float8, as illustrated in Fig. 5. It becomes immediately apparent that takum8 exhibits a markedly superior dynamic range in contrast to both posit8 and float8. Although the benchmark constants surpass the dynamic ranges of both posit8 and float8, takum8 encompasses all of these constants except Λ and M, which are narrowly missed but still adequately represented. Particularly noteworthy is the fact that, despite its expansive dynamic range, takum8 demonstrates the lowest minimum relative approximation error, surpassing both posit8 and float8. However, posit8 offers a slightly wider range of numbers with more than 0 fraction bits. This plot introduces the concept of the 'golden zone,' highlighted in yellow, which will be consistently referenced in all subsequent plots. The golden zone delineates the range of numbers wherein takum either matches or surpasses the performance of the reference IEEE 754 floating-point format. In this instance, the golden zone appears relatively narrow,

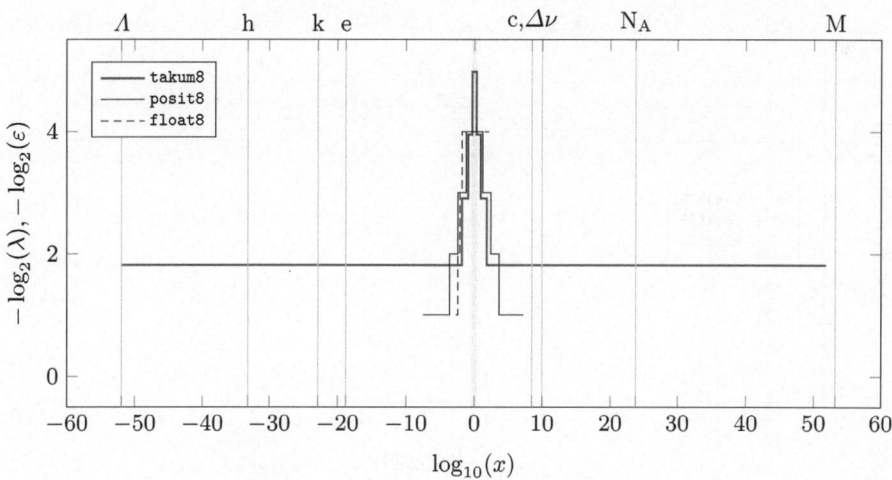

Fig. 5. Comparison of negative logarithmic relative approximation error upper bounds within an 8-bit budget for encoded numbers x. Higher values indicate lower relative error and, consequently, higher precision.

which is unsurprising considering that `float8` exhibits a very limited dynamic range coupled with relatively high overall precision. The presence of a discernible golden zone, despite the considerable dynamic range of `takum8`, underscores its commendable performance.

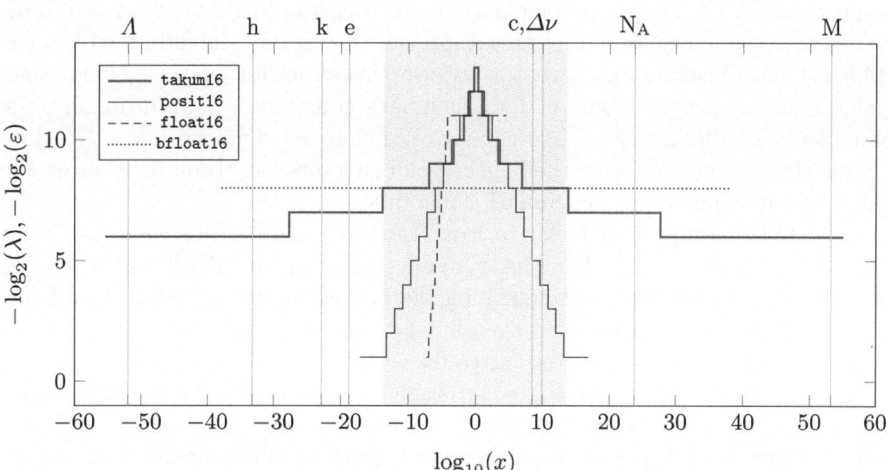

Fig. 6. Comparison of negative logarithmic relative approximation error upper bounds within a 16-bit budget for encoded numbers x. Higher values indicate lower relative error and, consequently, higher precision.

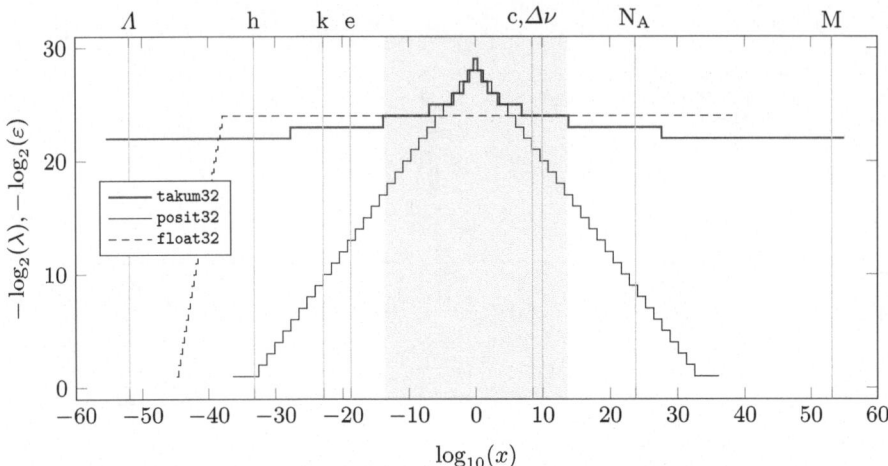

Fig. 7. Comparison of negative logarithmic relative approximation error upper bounds within a 32-bit budget for encoded numbers x. Higher values indicate lower relative error and, consequently, higher precision.

The comparison depicted in Fig. 6 introduces `bfloat16` as a fourth format alongside `takum16`, `posit16`, and `float16`. This addition is necessitated by the limited dynamic range of `float16`, which, among other reasons, served as the impetus for the development of `bfloat16` initially [26]. To ensure a fair evaluation of 16-bit floating-point formats, `bfloat16` is included as a benchmark against IEEE 754 floating-point numbers, notwithstanding its lack of standardisation. Subsequently, we adopt this approach throughout, including within the golden zone. Observations reveal a noteworthy extension of the golden zone, which now encompasses one of the benchmark constants (c). Additionally, the plot effectively illustrates the inadequacy of `bfloat16`'s dynamic range as it does not nearly encompass two of the benchmark constants (see Table 6). Conversely, `takum16` encompasses all benchmark constants.

The 32-bit comparison (refer to Fig. 7) exhibits similarities with the 16-bit comparison in the sense that the dynamic range of the IEEE 754 reference remains unchanged. The distinguishing factor lies in the inclusion of subnormals in `float32` and a reduced overall relative approximation error across all formats owing to the augmented bit count.

For the 64-bit plot (see Fig. 8), it is evident that the golden zone spans the entire dynamic range of `takum64`, offering a significant advantage over and compatibility with `float64`. In contrast, `posit64` sacrifices precision across its dynamic range, whereas `takum64` maintains at least more than 53 binary orders of magnitude of precision consistently.

Overall, takums offer a sufficiently broad dynamic range, and their relative accuracy exceeds that of posits near unity, nearly aligning with it for slightly larger exponents. Beyond this threshold, posits experience a rapid decline in

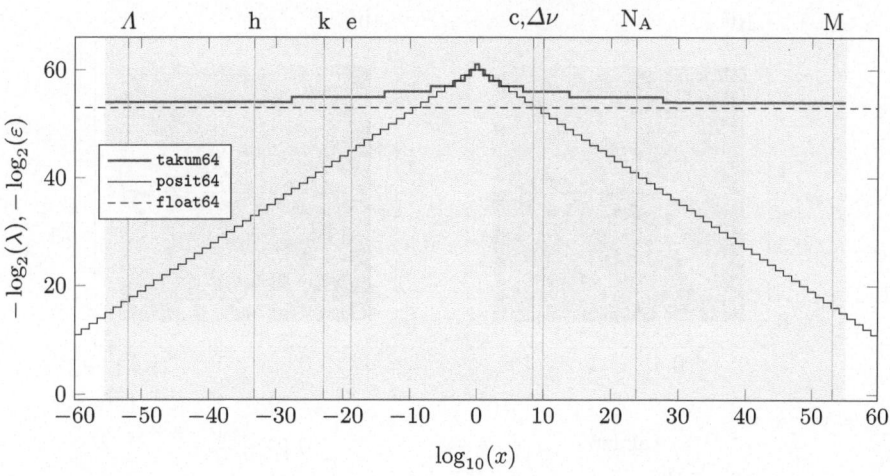

Fig. 8. Comparison of negative logarithmic relative approximation error upper bounds within a 64-bit budget for encoded numbers x. Higher values indicate lower relative error and, consequently, higher precision.

precision, whereas takums maintain a high level of relative accuracy, typically deviating by at most 1–2 binary orders of magnitude from, or even surpassing at higher bit counts, the respective reference IEEE 754 floating-point format.

5.6 Unary and Binary Arithmetic Closure

Besides the capability of a numerical format to represent a range of numbers, another crucial consideration is its efficacy in accurately representing the outcomes of arithmetic operations. The greater the number of arithmetic results that can be precisely represented, the lesser the accumulation of rounding errors across successive arithmetic operations. Considering x and y as numbers within a numerical system, round() as the associated rounding operation, and $\star$ as an arithmetic operation (such as addition, subtraction, multiplication, or division), we are concerned with both the absolute error

$$e_{\mathrm{abs}} := \mathrm{round}(x \star y) - (x \star y), \tag{99}$$

and the relative error

$$e_{\mathrm{rel}} := \frac{\mathrm{round}(x \star y) - (x \star y)}{x \star y}. \tag{100}$$

In order to condense the errors, which span a broad dynamic range, we introduce a clamped and logarithmically scaled 'relative precision' in binary magnitudes as

$$\eta(e_{\mathrm{rel}}) := \log_2(\max(1, -\log_2(|e_{\mathrm{rel}}|))), \tag{101}$$

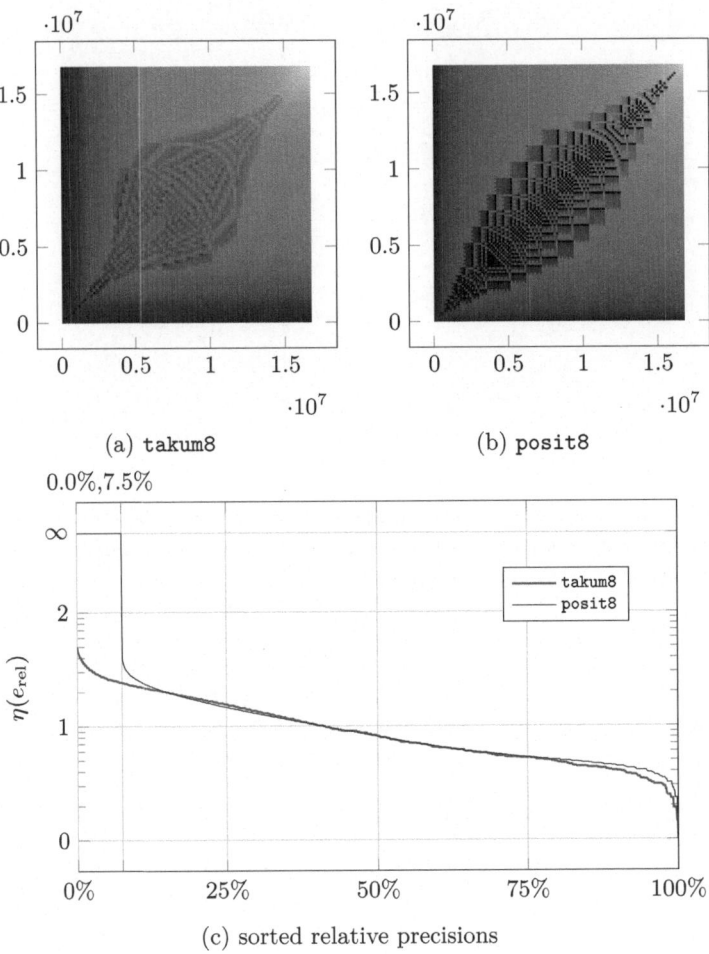

(a) takum8   (b) posit8

(c) sorted relative precisions

Fig. 9. Addition closure analysis of takum8 versus posit8 within posit8's positive dynamic range $[2^{-24}, 2^{24}] \approx [6.0 \times 10^{-8}, 1.7 \times 10^7]$.

which approaches ∞ when the relative error is zero and is bounded from below by zero, signifying a relative error greater than or equal to 10%.

For small bit string lengths, such as $n = 8$, it is feasible to compute the relative error for every possible arithmetic operation and present it visually in a two-dimensional plot. In this plot, the x-axis corresponds to all values of x, while the y-axis corresponds to all values of y (refer to Fig. 9 for an illustrative example). Each point (x, y) on the plot is colour-coded based on the rounded outcome of the arithmetic operation $x \star y$: negative relative errors are shaded in blue, zero errors in black, and positive relative errors in red. Blue and red hues are scaled in a perceptually uniform manner with respect to lightness, signifying that an increase in error magnitude is represented by enhanced brightness, irrespective of the direction.

To evaluate performance beyond mere visualisation, one can flatten the two-dimensional relative error 'matrix' into an array, compute the relative precision, and then sort the array in descending order. By plotting this sorted array with the x-axis in percent units, the ratio of exact operations within the number system becomes immediately apparent (as depicted in Fig. 9, where the exact operation ratios are 0.0% and 7.5% respectively). Both the concept of closure plots and the idea of sorting errors were adapted from [10], while the notion of relative precision is introduced here to enable the visualization of results in the plots, with higher values indicating better precision as opposed to worse.

New in this work is the consideration of the dynamic range inherent in various number formats. Rather than indiscriminately encompassing all numbers within each format, we identify the smallest common denominator in dynamic range and confine all number formats to this scope, encompassing solely positive numbers. This standardisation ensures uniformity of dynamic range across all formats. The inclusion of negative numbers is deemed superfluous when conducting a comprehensive comparison of addition and subtraction, as well as multiplication and division.

It is important to note that the approach of constraining the dynamic range places takums at a considerable disadvantage, given the format's expansive dynamic range. However, this method yields valuable insights into the feasibility of utilising takums within the same low-bit regimes for which posits have demonstrated efficacy, as evidenced in Sect. 5.5, where we have established the suitability of takums for general-purpose arithmetic based on their dynamic range and relative approximation error. For $n = 8$, we adopt the dynamic range of `posit8` as $[2^{-24}, 2^{24}] \approx [6.0 \times 10^{-8}, 1.7 \times 10^{7}]$, and for $n = 16$, we employ the dynamic range of `posit16` as $[2^{-56}, 2^{56}] \approx [1.4 \times 10^{-17}, 7.2 \times 10^{16}]$.

An additional novel aspect, refining the approach presented in [10], is the extension of our analysis to $n = 16$ bits, allowing for a direct comparison with `bfloat16`, a comparison not previously conducted.

In terms of addition, the impact of the logarithmic significand is evident in Fig. 9, where, for 8 bits, only 7.5% of `posit8` additions yield exact results, compared to a mere 0.0% for `takum8`. Despite differences in the frequency of exact operations, the sorted relative precisions between the two formats are strikingly similar. This observation persists in the context of 16 bits, as depicted in Fig. 10, where `bfloat16` has been incorporated into the comparison alongside `takum16` and `posit16`. Notably, `posit16` exhibits the highest rate of exact additions at 8.5%, while `bfloat16` fares considerably poorer with only 1.7% of additions resulting in exact outcomes. With a mere 0.0% of additions being exact, `takum16` occupies the third position. However, in non-exact cases, `takum16` outperforms `posit16`, displaying a higher relative precision, while both formats are behind `bfloat16` in this regard.

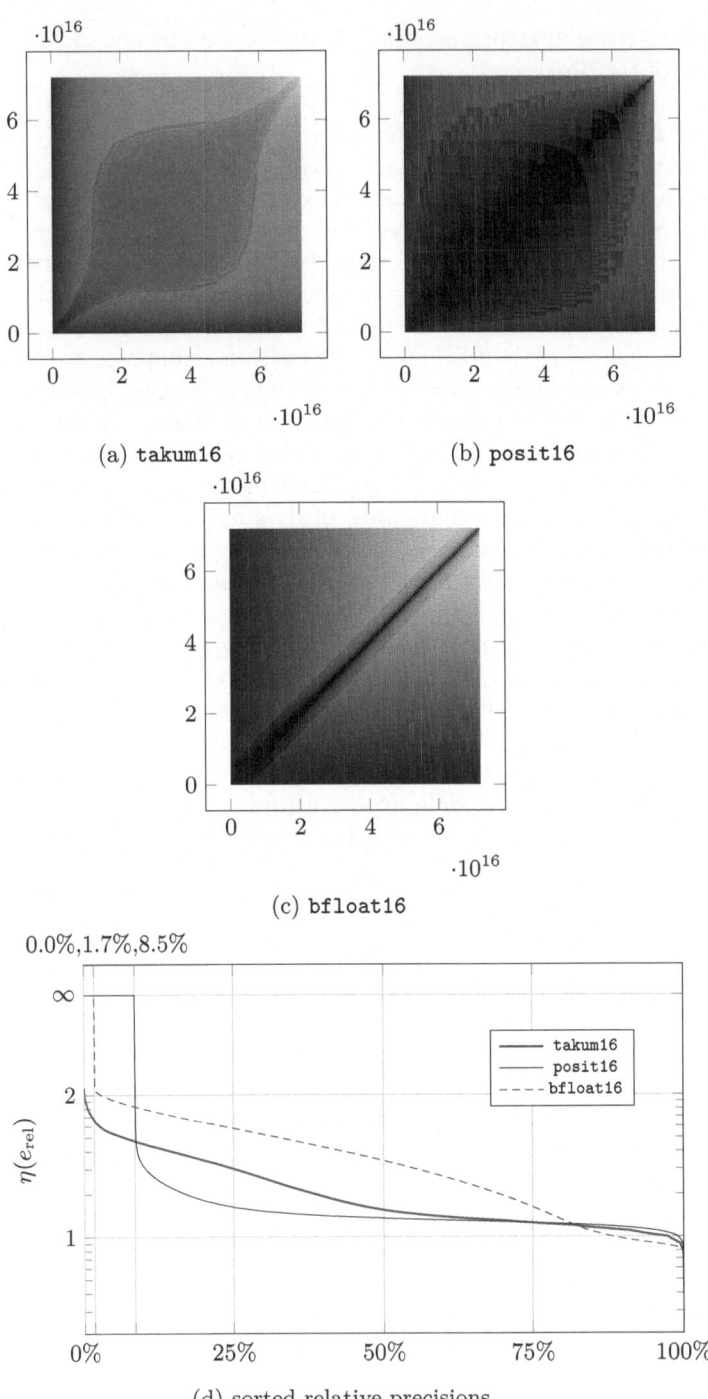

(a) `takum16`

(b) `posit16`

(c) `bfloat16`

(d) sorted relative precisions

Fig. 10. Addition closure analysis of `takum16` versus `posit16` and `bfloat16` within `posit16`'s positive dynamic range $\left[2^{-56}, 2^{56}\right] \approx \left[1.4 \times 10^{-17}, 7.2 \times 10^{16}\right]$.

For further insights into the evaluations of subtraction, readers are directed to Figs. 17 and 18 in the appendix, which present evaluations for 8- and 16-bit scenarios, yielding comparable results.

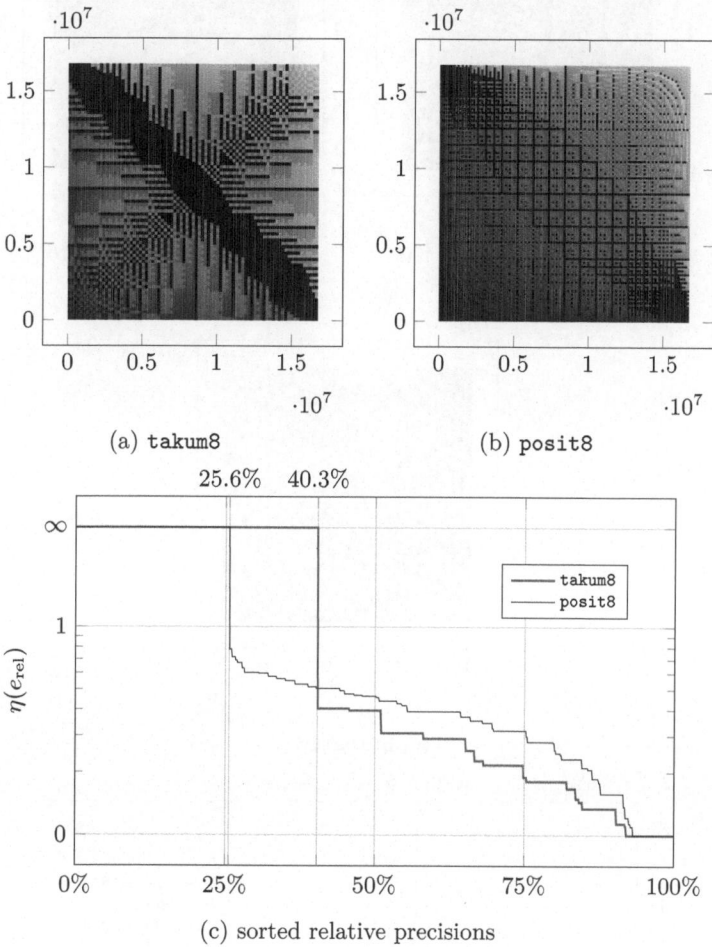

(a) takum8   (b) posit8

(c) sorted relative precisions

Fig. 11. Multiplication closure analysis of takum8 versus posit8 within posit8's positive dynamic range $\left[2^{-24}, 2^{24}\right] \approx \left[6.0 \times 10^{-8}, 1.7 \times 10^{7}\right]$.

In terms of multiplication, as illustrated for 8 bits in Fig. 11, it becomes evident that the logarithmic significand confers a distinct advantage. Specifically, takum8 exhibits a noteworthy 40.3% precision in exact multiplications, in stark contrast to posit8's mere 25.6%. However, the precision for non-exact multiplications marginally diminishes for takum8, a consequence attributable to its larger dynamic range, thus resulting in a higher density of posit8 when compared to takum8 in the common dynamic range used in this analysis. Mul-

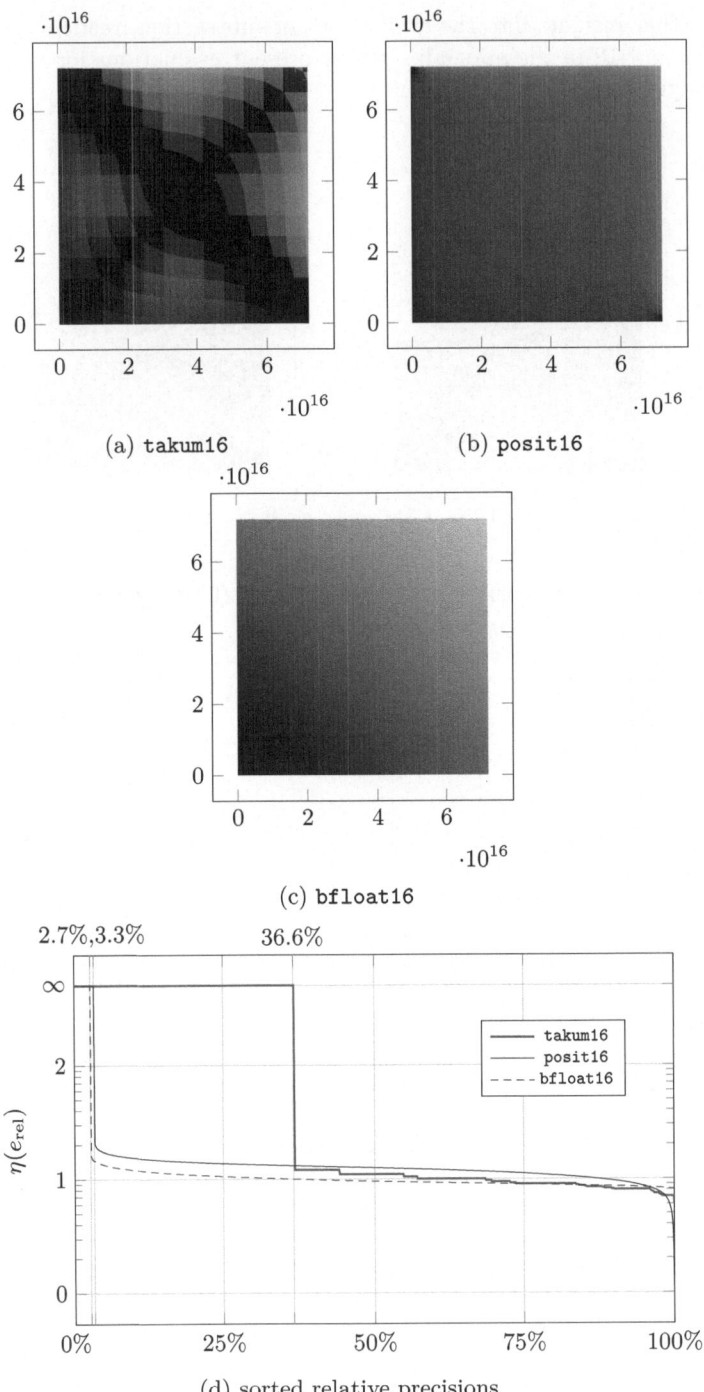

(a) `takum16`

(b) `posit16`

(c) `bfloat16`

(d) sorted relative precisions

Fig. 12. Multiplication closure analysis of `takum16` versus `posit16` and `bfloat16` within `posit16`'s positive dynamic range $\left[2^{-56}, 2^{56}\right] \approx \left[1.4 \times 10^{-17}, 7.2 \times 10^{16}\right]$.

tiplication in 16 bits (see Fig. 12) yields comparable results, but an even stronger dichotomy with 36.6% exact results for `takum16` versus only 2.7% and 3.3% exact results for `bfloat16` and `posit16` respectively. Refer to Figs. 19 and 20 in the appendix for 8- and 16-bit evaluations of division, which yielded similar results.

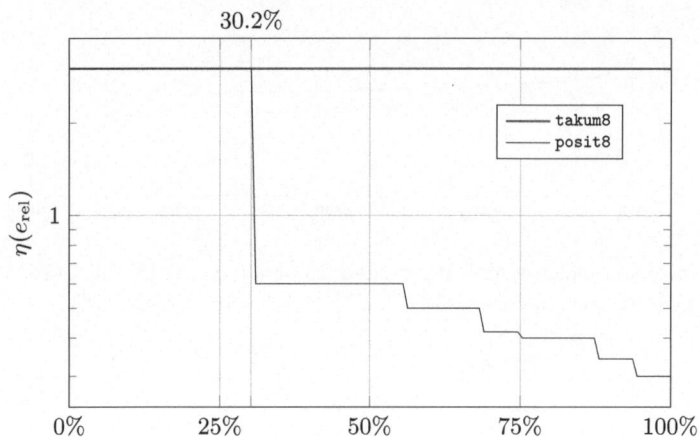

Fig. 13. Inversion closure analysis of `takum8` versus `posit8` within `posit8`'s positive dynamic range $[2^{-24}, 2^{24}] \approx [6.0 \times 10^{-8}, 1.7 \times 10^{7}]$.

As with binary operations, unary operations are also subject to analysis. Takums, by design, exhibit closure under inversion (refer to Proposition 7), ensuring that 100% of inversions yield exact results. This property distinguishes takums from posits, which lack such closure. For instance, in the case of $n = 8$ (as depicted in Fig. 13), it is observed that only a still acceptable 30.2% of `posit8` inversions yield exact results. Here we observe that extending to $n = 16$ bits (see Fig. 14) proves advantageous, averting erroneous conclusions, as in regard to the ratio of exact inversions `posit16` drops to a mere 0.3% at this juncture. Similar findings apply to `bfloat16`, with inversions exhibiting only 0.8% precision, thereby demonstrating a pronounced advantage for takums, particularly evident for larger values of n.

Another ubiquitous unary operation is the calculation of the square root, as illustrated for $n = 8$ in Fig. 15. With `takum8`, approximately 58.8% of square roots yield exact results, contrasting sharply with the mere 20.6% precision achieved by `posit8`. Even though this may appear to present a favourable ratio for `posit8`, upon scrutinising $n = 16$ bits (refer to Fig. 16), it becomes evident that the exactness rate of `posit16` diminishes to 1.3%, with `bfloat16` faring only marginally better at 3.1%. In contrast, `takum16` achieves a square root exactness rate of 84.1% without sacrificing precision in non-exact instances. These promising outcomes are further substantiated by the analysis of the squaring operation, as depicted in Figs. 22 and 21 in the appendix.

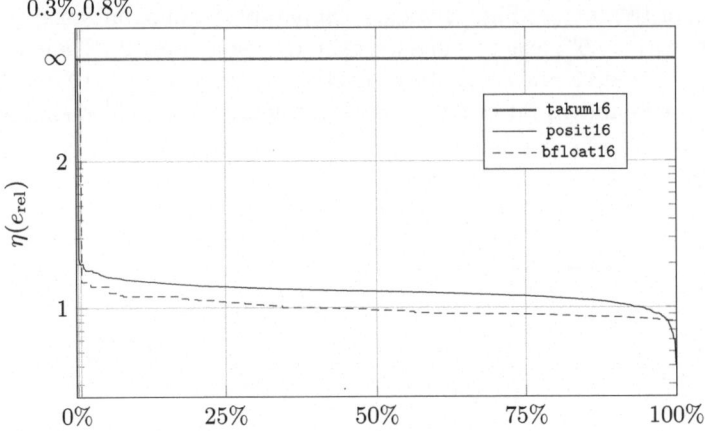

Fig. 14. Inversion closure analysis of `takum16` versus `posit16` and `bfloat16` within `posit16`'s positive dynamic range $[2^{-56}, 2^{56}] \approx [1.4 \times 10^{-17}, 7.2 \times 10^{16}]$.

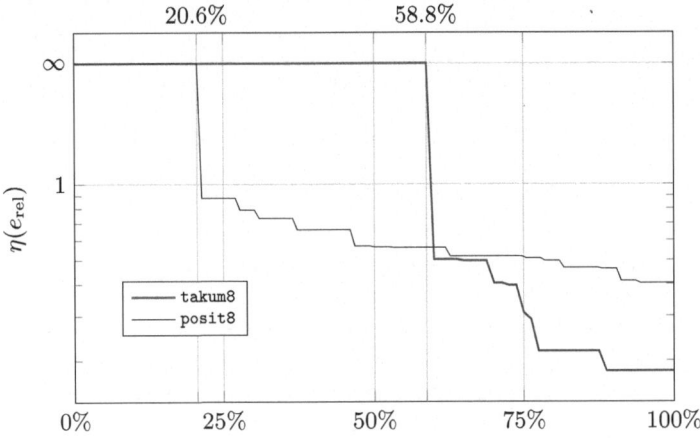

Fig. 15. Square root closure analysis of `takum8` versus `posit8` within `posit8`'s positive dynamic range $[2^{-24}, 2^{24}] \approx [6.0 \times 10^{-8}, 1.7 \times 10^{7}]$.

Overall, it is imperative for the reader to bear in mind that takums offer an almost 50% broader dynamic range than `bfloat16` and an even more substantially broader dynamic range than `posit8` and `posit16`. Despite this, takums consistently outperform or match posits and `bfloat16` in all scenarios except for addition and subtraction. The assessment of the advantage afforded by takums over posits and `bfloat16` is thus heavily contingent upon the specific application. Notwithstanding the drawbacks in the realm of addition and subtraction, the complete closure under inversion and markedly enhanced exactness in multiplication, division, square root, and squaring tilt the favour towards takums for general-purpose arithmetic.

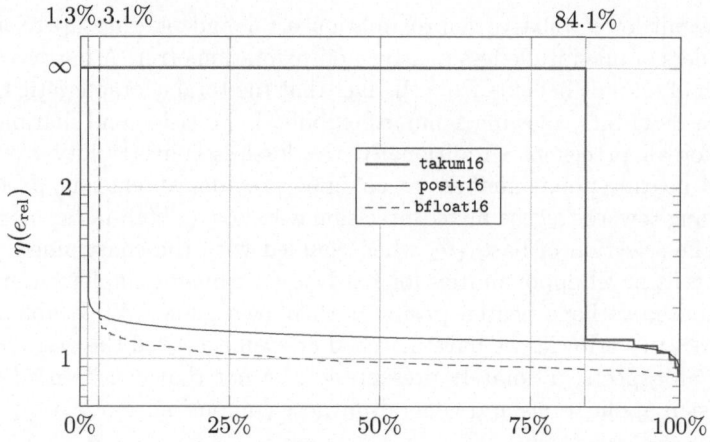

Fig. 16. Square root closure analysis of `takum16` versus `posit16` and `bfloat16` within `posit16`'s positive dynamic range $[2^{-56}, 2^{56}] \approx [1.4 \times 10^{-17}, 7.2 \times 10^{16}]$.

6   Conclusion

Upon expanding the GUSTAFSON criteria with two supplementary factors to encompass the desired dynamic range and establish a universally applicable one, we embarked on an exhaustive exploration of IEEE 754 floating-point numbers and posits, meticulously scrutinising their performance against these novel criteria.

The proposed and formally verified takum number format constitutes a logarithmic bounded-dynamic-range tapered precision number system that fulfils the GUSTAFSON and newly proposed dynamic range criteria, while demonstrating significantly enhanced coding efficiency for very high and very small magnitude numbers, and comparable numerical behaviour to posits. In comparison to posits, takums exhibit a constant dynamic range for all $n \geq 12$, with the characteristic length always bounded to 11 bits, ensuring $n - 12$ mantissa bits in all cases. These improvements are attributed to the absence of reliance on prefix codes, the bit-optimal leveraging of the unique properties of the number 255, and efficient utilisation of implicit bits for the characteristic bits. In essence, choosing a takum bit string length beyond 12 bits is solely driven by the pursuit of desired accuracy, rather than being a joint consideration of desired accuracy and dynamic range, as is the case with all other existing number formats. Based on the tapered format we have explored, we proposed the initiation of a potential new field termed 'tapered precision numerical analysis' as opposed to the established 'uniform precision numerical analysis'.

Regarding IEEE 754 floating-point numbers, the evaluation reveals that takums surpass the insufficient dynamic range of the 16 and 32 bit formats, offering significantly more precision for small exponents and only marginally reduced precision for larger exponents. Takums outperform `float64` by providing an overall lower relative approximation error and offering up to almost 8 binary orders of magnitude less relative approximation error. A hardware implementation of takums benefits from the fact that the total characteristic bit count never exceeds 11 bits, allowing genuine flexibility in precision and sharing of components for all precisions. Additionally, the lossless convertibility from 64-bit IEEE 754 floating-point numbers to `takum64` provides an efficient pathway for transitioning towards a true mixed-precision workflow in high-performance computing. The selection of base $\sqrt{e}$, when coupled with the constrained dynamic range, unveils novel opportunities for hardware arithmetic implementations.

In essence, we have beaten posits at their own game. While upholding all of GUSTAFSON's criteria, we have matched or even exceeded the original format in nearly all aspects, ultimately presenting a format that can be used both for low-precision applications and general-purpose computing.

References

1. Alam, S.A., Garland, J., Gregg, D.: Low-precision logarithmic number systems: Beyond base-2. ACM Trans. Arch. Code Optim. **18**(4), July 2021. https://doi.org/10.1145/3461699
2. Basir, M.S.S.M., Ismail, R.C., Naziri, S.Z.M.: An investigation of extended co-transformation using second-degree interpolation for logarithmic number system, pp. 59–63 (2020). https://doi.org/10.1109/FORTEI-ICEE50915.2020.9249931
3. Burks, A.W., Goldstine, H.H., Neumann, J.: Preliminary discussion of the logical design of an electronic computing instrument. In: Randell, B., et al. (eds.) The Origins of Digital Computers, pp. 399–413. Springer Berlin Heidelberg, Berlin, Heidelberg (1982). https://doi.org/10.1007/978-3-642-61812-3_32
4. Carlström, J.: Wheels - on division by zero. Math. Struct. Comput. Sci. **14**, 143–184 (2004). https://doi.org/10.1017/S0960129503004110
5. Coleman, J.N., Ismail, R.C.: Lns with co-transformation competes with floating-point. IEEE Trans. Comput. **65**(1), 136–146 (2016). https://doi.org/10.1109/TC.2015.2409059
6. Coleman, J.N., Chester, E.I., Softley, C.I., Kadlec, J.: Arithmetic on the european logarithmic microprocessor. IEEE Trans. Comput. **49**(7), 702–715 (2000). https://doi.org/10.1109/12.863040
7. De Dinechin, F., Forget, L., Muller, J.M., Uguen, Y.: Posits: the good, the bad and the ugly. In: CoNGA'19, New York, NY, USA, 2019. Association for Computing Machinery. https://doi.org/10.1145/3316279.3316285
8. John, L.: Gustafson. Posit arithmetic, Every bit counts (2024)
9. Gustafson, J.L.: The End of Error: Unum Computing. Chapman & Hall/CRC Computational Science. CRC Press, April 2015. ISBN 9781482239874
10. Gustafson, J.L., Yonemoto, I.T.: Beating floating point at its own game: Posit arithmetic. Supercomput. Front. Innov. **4**(2), 71–86 (2017). https://doi.org/10.14529/jsfi170206
11. Gustafson, J.L., et al.: Standard for posit^{TM} arithmetic (2022). https://web.archive.org/web/20220603115338/https://posithub.org/docs/posit_standard-2.pdf
12. IEEE standard for floating-point arithmetic July (2019)

13. Johnson, J.: Efficient, arbitrarily high precision hardware logarithmic arithmetic for linear algebra. In: 2020 IEEE 27th Symposium on Computer Arithmetic (ARITH), pp. 25–32, Los Alamitos, CA, USA, June (2020). IEEE Computer Society. https://doi.org/10.1109/ARITH48897.2020.00013

14. Johnson, J.: Rethinking floating point for deep learning, p. 1–8 (2018)

15. Kahan, W.: Lecture notes on the status of IEEE standard 754 for binary floating-point arithmetic. October 1997. https://web.archive.org/web/20240308034347, https://people.eecs.berkeley.edu/~kahan/ieee754status/IEEE754.PDF

16. Kahan, W.: Names for standardized floating-point formats (April 2002). https://web.archive.org/web/20231227155514, https://people.eecs.berkeley.edu/wkahan/ieee754status/Names.pdf

17. Kharya, P.: Tensorfloat-32 in the a100 gpu accelerates ai training, hpc up to 20x. May 2020. https://web.archive.org/web/20231126174430, https://blogs.nvidia.com/blog/tensorfloat-32-precision-format

18. Kouretas, I., Paliouras, V.: Logarithmic number system for deep learning. In: 2018 7th International Conference on Modern Circuits and Systems Technologies (MOCAST), pp. 1–4. IEEE (June 2018). https://doi.org/10.1109/MOCAST.2018.8376572

19. Lilley, C.: Color on the Web, chapter 16, page 271-291. John Wiley & Sons, Ltd, 2023. ISBN 9781119827214. https://doi.org/10.1002/9781119827214.ch16

20. Lindstrom, P., Lloyd, S., Hittinger, J.: Universal coding of the reals: alternatives to IEEE floating point (2018). https://doi.org/10.1145/3190339.3190344

21. Miyashita, D., Lee, E.H., Murmann, B.: Convolutional neural networks using logarithmic data representation, pp. 1–10, March (2016)

22. Morris, R.: Tapered floating point: a new floating-point representation. IEEE Trans. Comput. **C-20**(12), 1578–1579 (1971). https://doi.org/10.1109/T-C.1971.223174

23. Muller, J.M.: Discrete basis and computation of elementary functions. IEEE Trans. Comput. **34**(09), 857–862 (1985). ISSN 1557-9956. https://doi.org/10.1109/TC.1985.1676643

24. Ramachandran, A., et al.: Algorithm-hardware co-design of distribution-aware logarithmic-posit encodings for efficient DNN inference, pp. 1–6 (2024)

25. Quevedo, L.T.: Automática: Complemento de la teoría de las máquinas. *Revista de Obras Públicas* **(2043)**, 575–583, November (1914). https://quickclick.es/rop/pdf/publico/1914/1914_tomoI_2043_01.pdf

26. Wang, S., Kanwar, P.: Bfloat16: The secret to high performance on cloud tpus. August 2019. https://web.archive.org/web/20190826170119/https://cloud.google.com/blog/products/ai-machine-learning/bfloat16-the-secret-to-high-performance-on-cloud-tpus

Breaking New Ground in AI with Posit Arithmetic and Vision Transformer

Ashwini Jaya Kumar, Rajaraman Subramanian[(⊠)],
and Sundararajan Venkatachari

Calligo Technologies Pvt. Ltd., Bengaluru, India
{ashwini.jayakumar,rajaraman.subramanian,sundar.chari}@calligotech.com

Abstract. In the dynamic realm of artificial intelligence, the quest for more efficient and precise deep learning models stands as an enduring challenge. Low-bit encoding presents a transformative approach to neural network representation by reducing their precision to lower-bit formats, thereby substantially curtailing memory usage and computational demands.

Vision Transformer (ViT) have garnered substantial attention at the intersection of Transformers and Computer Vision. However, ViTs are computationally expensive when dealing with large-scale models and datasets, and encoding ViT with low-bit arithmetic has significant advantages in terms of computational efficiency and resource requirements.

This paper introduces a low-bit encoding of Vision Transformer using Posit Arithmetic, a novel number system known for its precision, reduced execution time, storage economy, and lower power consumption compared to traditional floating-point numbers.

We conducted a comparative study involving low-precision Posit empowered Stacked Convolutional Neural Networks (Stacked CNNs), conventional ViT, and Stacked Vision Transformer (Stacked ViT). Our findings reveal that the Stacked Vision Transformer competes effectively with the Stacked Convolutional Neural Network baseline with respect to training time and accuracy. This underscores the potential of low-bit encoding using Posit as a compelling solution for AI applications that must adhere to stringent computational constraints.

Through meticulous experimentation and comprehensive evaluation, this paper unveils the remarkable advantages of low-bit encoding, offering new insights into the future of efficient and accurate deep-learning models.

Keywords: Convolutional Neural Networks · Vision Transformer · Posit Arithmetic · Universal Library

1 Introduction

Deep neural networks (DNNs) have emerged as a transformative class of machine learning models, revolutionizing diverse applications across domains such as

M. Michalewicz et al. (Eds.): CoNGA 2024, LNCS 14666, pp. 52–73, 2024.
https://doi.org/10.1007/978-3-031-72709-2_2

image classification, speech recognition, and natural language processing. These networks excel at learning hierarchical representations of data, contributing to their effectiveness in solving complex tasks. However, the training of DNNs poses significant challenges due to the substantial memory and computational demands, particularly for large models.

An innovative approach to overcome the memory and computational complexity challenges is to train DNNs using reduced-precision data representation. It is commonly used in scenarios where the full precision of higher-bit representations is not necessary. By employing precision reduction techniques, such as low-bit encoding, we aim to strike a balance between computational efficiency and model performance. Utilizing fewer bits for data representation reduces the memory footprint, alleviating the strain on computational resources.

Posit Arithmetic, a revolutionary numerical system, is praised for its precision, optimized execution time, efficient storage, and reduced power consumption when compared to conventional floating-point numbers. Posit is good for applications like machine learning [9]. The advantages of applying 8-bit posit to the deep neural network are that it requires less number of bits, good accuracy, less power consumption, and less memory requirement. More recently, the posit format has shown promise over floating point with a larger dynamic range, higher accuracy, and better closure [1,5–7].

Vision Transformers (ViTs) have recently emerged as a promising approach to computer vision tasks like image classification, sparking a wave of research at the intersection of Transformers and Computer Vision (CV) [3]. ViT is a type of deep neural network architecture that leverages the self-attention mechanism used in Transformers [12] to model spatial dependencies in images. Unlike traditional Convolutional Neural Networks (CNNs), ViTs operate directly on the image's pixel space, without relying on handcrafted features. ViTs require significant computational resources, making them difficult to train and deploy in resource-constrained environments.

To address these challenges, researchers are exploring a variety of techniques, such as distillation and pruning, to make ViTs more efficient and practical. A low-bit encoded ViTs represents a promising and innovative avenue within the realm of computer vision research.

ViT is designed in a way such that there are no convolution layers and the image patches are given as input to the Transformers. We want to take advantage of both CNN and ViT architectures to capture local correlations at the pixel level and also the global dependencies. Hence we are modifying ViT a bit and considering the new architecture as Stacked Vision Transformer (Stacked ViT). In simple words, it is adding a convolution stack at the input of ViT instead of image patches. We are adding convolution stacks at the input of both CNN and ViT and they are called Stacked CNN (Fig. 8) and Stacked ViT (Fig. 10). We are making a thorough comparison of performance between Stacked CNN, ViT, and Stacked ViT with low-bit encoding.

In this paper, IEEE single precision and double precision floating point number [17] is represented as Float32 and Float64 respectively. 16-bit Brain floating point number [4] is represented as Bfloat16. The notation to denote posits of bit width n and es bits reserved for the exponent is $< n, es >$ [1]. We are using $n = 8$ bits with exponent $es = 0$ i.e., $< 8, 0 >$ or Posit8 or P8. Figure 1 shows the space occupied by an image (shown in Fig. 2) in Kilobytes (kbytes) using Float64, Float32, Bfloat16 and Posit8. Posit8 shows the least space occupied by an image. The comparison of an image encoded with Float32, Bfloat16 and Posit8 is shown in Fig. 2. There is no visible difference between the three datatypes. When we compute the absolute pixel difference between Float64 (python default) and Float32, Bfloat16 and Float32, and Posit8 and Float32, we can notice that residue is more between Posit8 and Float32 and is shown in Fig. 3. We present a thorough analysis of loss of precision while using Posit8 in Sect. 3 and show that the residual error is negligible and doesn't affect the training accuracy.

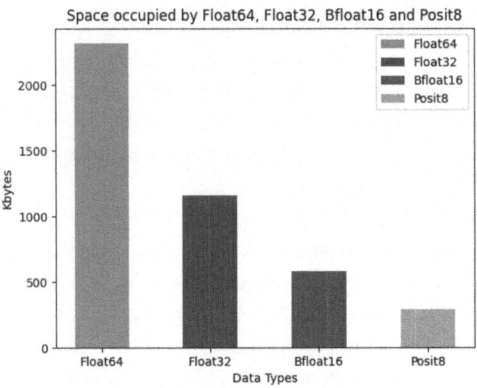

Fig. 1. Space occupied by Float64, Float32, Bfloat16 and Posit8 for an image in Fig. 2.

Fig. 2. Float32, Bfloat16 and Posit8 encoded images

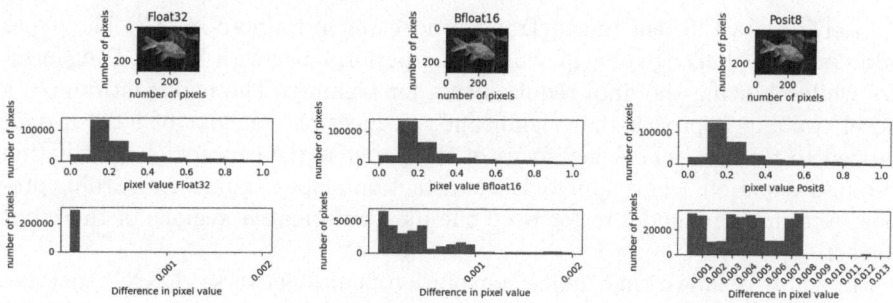

Fig. 3. Top row contains images encoded with Float32, Bfloat16 and Posit8 format. The second row represents the histogram of the pixels in an image. The third row illustrates the difference (absolute value) between pixel values, between (Float32 and Python default (Float64)), (Bfloat16 and Float32), and, (Posit8 and Float32).

The main contributions of this paper are:

- We have shown that posit can be a direct drop-in replacement for both Float32 and Bfloat16 as the benefits of low-bits like posits are unmatchable.
- Demonstrate that low-bit posit, as low as 8-bit can be used in convolutional neural networks and vision transformer architecture without losing accuracy.
- Stacked ViT, which makes use of both CNN and ViT, outperforms ViT and Stacked CNN in terms of accuracy and time.

The background is explained in Sect. 2. Qualitative analysis of low bit encoding using posit arithmetic is presented in Sect. 3. Posit encoding of nd Stacked ViT are explained in Sect. 4. Implementation is explained in Sect. 5. Results and Discussion are presented in Sect. 6. The Conclusion is presented in Sect. 7.

2 Background

2.1 Vision Transformer

It is established that Vision Transformers (ViTs) outperform Convolutional Neural Networks (CNNs) [3] in computer vision applications like image classification. Posit arithmetic can be incorporated successfully for image classification [5] and Generative Adversarial Network [2]. In this work, we are exploring the behavior of Vision Transformer when we encode the network with low-bit posit [1] and compare that with the CNN baseline.

As research on ViTs progresses, various variants and modifications of the original architecture have been proposed. Models like the Swin Transformer [26] explore different strategies for processing image patches or scaling to larger datasets. It introduces the concept of "shifted windows" to process image patches efficiently, reducing computational costs. Also, it achieves competitive results in various computer vision tasks.

DeiT (Data-Efficient Image Transformer) was introduced in [27]. The paper addresses the challenge of achieving strong performance with Vision Transformers while reducing the data requirements for training. The paper introduces a novel training approach that significantly reduces the amount of labeled data needed to train a ViT. A key focus of the paper is the concept of "distillation through attention." This approach uses attention maps to guide the learning process, allowing the model to focus on the most informative regions of the input image.

In [23] it is shown that unlike convolution neural networks (CNNs) that can be improved by stacking more convolutional layers, the performance of ViTs saturates fast when scaled to be deeper. In [24], a hybrid network is built by taking advantage of both CNNs and transformers, following the design in modern CNNs (e.g., ResNet [25]).

2.2   Posit Arithmetic

Deep neural networks (DNNs) have revolutionized the field of machine learning by achieving remarkable accuracy in various tasks. However, the computational intensity of DNNs during training and inference poses significant challenges in terms of energy consumption and speed. To address these challenges, [28] explores the application of posits, a versatile numerical representation, for fast approximations of activation functions within DNNs. [29] introduces the concept of unums, from which posits are derived. It presents the advantages of unum arithmetic over traditional fixed-point and floating-point representations.

The posit number format has a distinctive property compared to other formats which results in better numerical stability in many application domains. The distribution of representable values in posits is more concentrated to a central point in the log2 domain [10]. This property will benefit certain applications where most of the values are concentrated to a specific range. Complete details of the posit format and its related arithmetic can be found in the posit standard [11]. The numerical stability of posit to solve linear systems is evaluated in [22]. A thorough investigation of posit versus float is presented in [9].

In [8], through a comprehensive analysis of quantization with various data formats, it is demonstrated that the posit format shows great potential to be employed in the training of DNNs. A comprehensive comparison among posit, float, and fixed-point is presented. From multiple perspectives, including decimal accuracy, mean related error, and mean absolute error, the outstanding representation ability of posit is well justified. In [30], a thorough analysis of the application of posit to the deep neural networks is presented.

2.3   Number System

IEEE 754 Single Precision. IEEE Single precision floating point number (Float32) is represented as in Eq. 1. It decomposes a number into a sign (1-bit), exponent (8-bits) and mantissa (23-bits):

$$X = (-1)^s * (2)^{exponent-127} * mantissa \tag{1}$$

1 bit	8 bits	23 bits
S(sign)	E(Exponent)	M(Mantissa)

Fig. 4. IEEE Single Precision Floating Point Number System

Representation of the Single Precision Floating Point Number System is shown in Fig. 4.

IEEE standard for floating-point arithmetic is the most common implementation that modern computing systems have adopted. They have several shortcomings: 1) Identical computations can lead to multiple results across different computing platforms(different computers). 2) Multiple-bit patterns are used for handling exceptions such as the NaN. 3) Exponents are usually too large and not adjustable

To overcome the disadvantages of Floating point numbers, in 2017, a posit number representation system [1,20], is introduced. It claims that it provides more accurate answers with an equal or smaller number of bits and simpler hardware. Posit number system is explained in the next section.

Posit Number System. The numerical value of a posit numer X is given by Eq. 2 and explained in [6],

$$X = (-1)^s * (useed)^k * 2^e * (1 + f) \tag{2}$$

where s is the sign bit, $useed = 2^{2^{es}}$, k is the regime value, e is the unsigned exponent (if $es > 0$), and f is the mantissa of the number. We use the notation $< n, es >$ to denote posits of bit width n and es bits reserved for the exponent, respectively. Posit Number system with sign, regime, exponent and fraction bits is shown in Fig. 5.

1 bit	$k + 1$ bits	e bits	f bits
S(sign)	R(Regime)	E(Exponent)	F(Fraction)

Fig. 5. Posit Number System

We have considered posit $< 8, 0 >$ in this work. The posit number system has many interesting properties, such as lack of underflow or overflow, less computation time, low bits, good accuracy, and a flexible number of bits and exponent configuration. If an application can switch from using 64-bit IEEE floats to using 32-bit posits, for example, it can fit twice as many numbers in memory at a time. More information on posit is available in [19].

Bfloat16 Number System. Bfloat16 is a custom 16-bit floating point format for machine learning that's comprised of one sign bit, eight exponent bits, and seven mantissa bits. It is designed keeping deep learning applications in mind [4]. It is designed to reduce memory and storage requirements while providing good numerical accuracy compared to other low-precision formats. It is equivalent to a standard single-precision floating-point value with a truncated mantissa field. Representation of Bfloat16 is shown in Fig. 6.

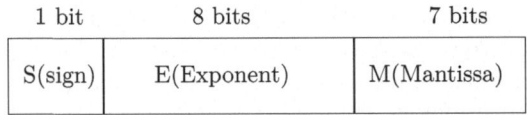

Fig. 6. Bfloat16 Number System

The key thing here is that for machine learning operations you don't need the levels of precision in terms of binary powers that other calculations might require. But you do want speed of operation, and that's what bfloat16 is aimed at [4].

3   Qualitative Analysis of Low Bit Encoding Using Posit Arithmetic

3.1   Loss of Precision Analysis

In low-bit encoding, some level of precision may be lost during mapping floating-point to low-bit values. Measuring the loss of precision when performing data compression or quantization, such as in the case of using low-bit encoding or other compression techniques, can be critical to understanding the impact on the data. We are considering Mean Squared Error (MSE) and Mean Relative Error (RE) for our analysis.

Mean Squared Error (MSE): MSE is a widely used metric to measure the loss in low-bit encoding algorithm. It is the average squared difference between the original (uncompressed) data and the compressed (low-bit encoded) data. A higher MSE indicates a greater loss of precision. It is calculated as in Eq. 3:

$$MSE = \frac{1}{N} \sum_{i=1}^{N} (x_i - \hat{x}_i)^2 \qquad (3)$$

where, $\frac{1}{N}$ represents the reciprocal of the number of data points, denoted as N, and it's used to calculate the sum of squared differences for each data point, i is the index used in the summation, ranging from 1 to N, x_i represents the original data point, $\hat{x}_i$ represents the low-bit encoded data point.

Table 1 shows the loss of precision using MSE analysis. The loss is comparatively more between Float32 and Posit8, in the order of $1e - 05$, which is negligible.

Table 1. Mean Squared Error and Relative Error Analysis for an image shown in Fig. 2

Data Type	MSE	MRE (%)
(Float64, Float32)	8.034892370517e−17	2.516954621615e−06
(Float32, Bfloat16)	3.450959996394e−07	0.164948578458279
(Float32, Posit8)	2.0878807054768e−05	1.5212502246412363

Mean Relative Error (MRE): The MRE (ε_r) is a measure of the accuracy of an approximation compared to the true value. It's expressed as a ratio of the absolute error to the absolute value of the true value as shown in Eq. 4:

$$\varepsilon_r = \frac{\sum |TrueValue - Low\_Precision\_Value|}{\sum |TrueValue|} * 100 \tag{4}$$

where, True Value is the actual value. Low-precision value is the calculated or estimated value. The vertical bars (||) denote the absolute value, ensuring that both the numerator and the denominator are positive.

Table 1 shows the loss of precision using Mean Relative Error (MRE) analysis. The mean relative error quantifies the average percentage difference between the low-bit encoded values and the actual values. A lower MRE indicates that the predictions are, on average, closer to the actual values, which is a desirable outcome. In Table 1, MRE is 1.52% between Float32 and Posit8 indicating an increased level of error compared to FLoat64 and Float32, and Float32 and Bfloat16. It should be noted that even with high MRE, the impact of Posit8 on the final accuracy is negligible compared to Float32.

3.2   Loss of Image Quality Analysis

The edges of an image form the most important feature of an image for various computer vision applications like image classification and object detection. We are estimating the loss of image quality by computing edge density analysis before and after Posit8 conversion.

Edge Density Analysis (EDA). Edge density is a measure of the presence of edges or transitions in an image. It quantifies the number of edges or transitions per unit area. One common approach to measure edge density is by calculating the gradient magnitude of the image using techniques like the Sobel operator or the Canny edge detector and then counting the number of edges. We are using a Canny edge detector here. Edge density is given by Eq. 5:

$$\text{Edge Density} = \frac{\text{Number of Edge Pixels}}{\text{Image Width} \times \text{Image Height}} \tag{5}$$

Table 2. Edge Density Analysis for the images shown in Fig. 2

Data Type	EDA
Float32	13.798438637650676
Bfloat16	13.798438637650676
Posit8	13.415176765633849

A high edge density suggests that there are many distinct edges or boundaries in an image. A low edge density means that there are fewer distinct edges or boundaries in the image. In Table 2, we can see that EDA is slightly reduced for Posit8, which implies that there is a slight loss of information.

4   Posit Encoded Stacked Convolutional Neural Network, Vision Transformer and Stacked Vision Transformer

Posit encoding aims to leverage the benefits of Posit numerical representation, such as reduced memory requirements and computational complexity. Employing Posit encoding in both input and weights enhances the precision and efficiency of the DNN model, potentially providing advantages in terms of accuracy and resource utilization.

Algorithm 1. Low-Bit Encoding using Posits

Input: Input sequence X, desired bit precision n and exponent es
Output: Encoded input X_{encoded}
for each value x in X **do**
    $X_{\text{encoded}} \leftarrow \text{Map}(x, Posit < n, es >)$
end for
return X_{encoded}

4.1 Posit Encoding

The encoding process often includes mapping floating-point values to a lower-precision Posit Format. The mapped low-precision function can be represented as in Eq. 6.

$$P_i = \text{Mapping}(X_i, \text{Posit Format}) \tag{6}$$

Every input pixel and initial weight (parameter) of the neural network is encoded into a posit value. An algorithm to encode input and weight values to posit is shown in Algorithm 4.

4.2 Convolution Stack

The architecture of a neural network is very critical for the overall performance. We generally use convolutional neural networks with pooling and fully connected layers. The first convolutional layer is designed to capture all global features and as we top-down, by adding more layers, we try to capture intrinsic features in an image for example [21]. The function of pooling is to progressively reduce the spatial size of the representation to reduce the number of parameters and computations in the network. The most commonly used pooling is max-pooling. We have the freedom to add any number of CNN and max-pooling layers but decision has to be made depending on the database and the application. In addition to max-pooling, we are adding batch normalization in order to deal with internal covariant shift [16].

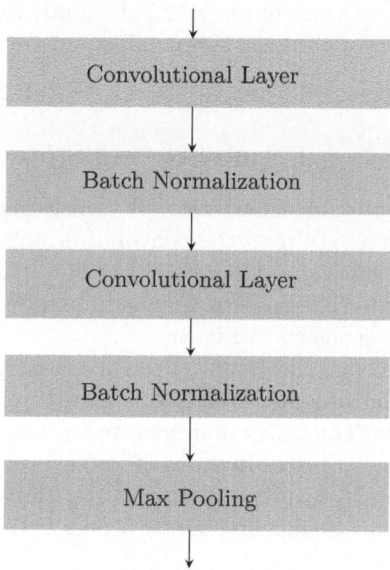

Fig. 7. Convolution Stack

We have used a convolutional neural network consisting of a convolution stack, each stack contains: 1) Convolutional Layer 2) Batch Normalization 3) Convolutional Layer 4) Batch Normalization, and 5) Max pooling Layer. The convolution stack is shown in Fig. 7.

Convolution Layer: Let l be the convolutional layer. Then the input of layer l comprises $m_1^{(l-1)}$ feature maps from the previous layer, each of size $m_2^{(l-1)} \times m_3^{(l-1)}$ .In the case where $l = 1$, the input is a single image I consisting of one or more channels. This way, a convolutional neural network directly accepts raw images as input. The output of layer l consists of m_1^l feature maps of size $m_2^l \times m_3^l$. The ith feature map in layer l denoted $Y_i^{(l)}$ is computed as

$$Y_i^{(l)} = B_i^{(l)} + \sum_{j=1}^{m_1^{(l-1)}} K_{i,j}^{(l)} * Y_j^{(l-1)} \tag{7}$$

where $B_i^{(l)}$ is a bias matrix, $k_{i,j}^{(l)}$ is the filter of size $2h_1^{(l)} + 1 \times 2h_2^{(l)} + 1$ connecting the jth feature map in $(l-1)$ layer with i^{th} feature map in layer l

Batch Normalization: Consider we have d number of hidden units in a hidden layer of any deep neural network. We can represent the activation values of this layer as $x = [x_1, x_2, \ldots x_d]$. Now we can normalize the kth hidden unit activation using the formula below in Eq. 8.

$$\hat{x} = \frac{x^k - E(x^k)}{\sqrt{var(x^k)}} \tag{8}$$

where, $\hat{x}$ the normalized value of the kth hidden unit, $E(x^k)$ is the expectation of the kth units values also called the mean value and $var(x^k)$ is the variance of the kth hidden unit.

4.3   Stacked Convolutional Neural Network (Stacked CNN)

We have used a convolutional neural network consisting of three convolution stacks. The input image is given to the convolution layer of the convolutional stack. Finally, the third stack's max-pooling layer output is given to a fully connected layer with dropouts. The Stacked Convolutional Neural Network (Stacked CNN) used in this work is shown in Fig. 8.

The stacking of convolutional layers allows a hierarchical decomposition of the input. The filters that operate directly on the raw pixel values will learn to extract low-level features. The filters that operate on the output of the first layer may extract features that are combinations of lower-level features. This process continues until very deep layers are extracting faces, animals, houses, and so on. The abstraction of features to high and higher orders as the depth of the network is increased.

We have used Universal Number Library [18] to convert float into Posit < 8, 0 >. The input image array is first converted to Posit and then weights are initialized in the first pass of the network training, the initialized weights are

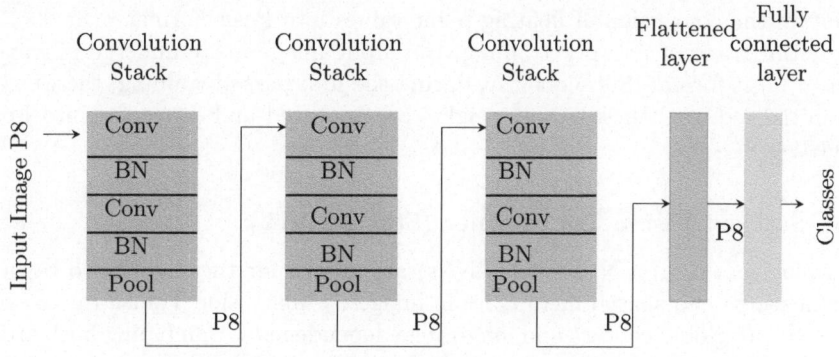

Fig. 8. Stacked Convolutional Neural Network (Stacked CNN)

captured and converted to Posit values. The $P8$ symbol in Fig. 8 shows the flow of posit value in the network.

4.4 Vision Transformer

Vision Transformer (ViT) [3] consists of 1) Input Image patches, 2) Transformer, 3) Multi-Layer Perceptron, and 4) Output classes. It splits each image into a sequence of tokens (i.e. non-overlapping patches) with fixed length and then passed to the transformer layer, consisting of, Layer Normalization (LN), Multi-Head Self-Attention module (MHSA) and Positionwise Feed-forward module or Multi-Layer Perceptron (MLP), to model these tokens [3,12]. Transformers exclusively rely on the self-attention mechanism to capture global dependencies. Vision Transformer is shown in Fig. 9.

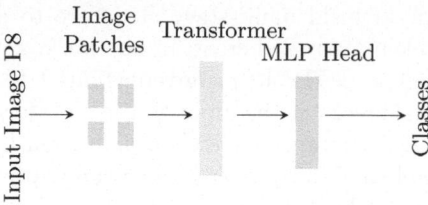

Fig. 9. Vision Transformer (ViT)

Vision Transformer has much less image-specific inductive bias than CNNs. In CNNs, locality, two-dimensional neighborhood structure, and translation equivariance are combined into each layer throughout the whole model. In ViT, only MLP layers are local and translationally equivariant, while the self-attention layers are global.

The images which are divided into non-overlapping image patches are encoded into Posit8 or P8. We utilized the Universal Number Library [18] to facilitate the conversion of floating-point values into Posit Format $< 8, 0 >$. In the initial stage of network training, the input image array undergoes conversion to Posit format. Subsequently, during the first pass of training, the weights are initialized, and these initial weights are captured and converted into Posit values.

4.5 Stacked Vision Transformer (Stacked ViT)

Convolutional Neural Networks (CNNs) are known for their ability to capture local features and spatial hierarchies in images, while Vision Transformers excel at capturing global context and long-range dependencies. Combining both architectures allows the model to benefit from local and global information, potentially improving the understanding of complex visual patterns.

We are taking advantage of both CNN and ViT architecture to design our Stacked ViT. Our proposed neural network architecture consists of three convolutional stacks cascaded with ViT. It is called Stacked Vision Transformer (Stacked ViT) as we are concatenating Stacked CNN with ViT. The outer layer of the convolutional stack is a max-pooling layer, the output of the max-pooling layer is given as input to the transformer. Basically, the input of the transformer is a projection of an image on a lower dimensional space. The Stacked Vision Transformer (Stacked ViT) is shown in Fig. 10. We need a convolutional layer with a transformer to capture local relationships. Images have a strong 2D local structure: spatially neighboring pixels are usually highly correlated. The CNN architecture forces the capture of this local structure by using local receptive fields, shared weights and, spatial subsampling.

We have a variant of vision transformer for small size data set [13, 14]. Convolution with Vision Transformer is not new. In [15], convolution is required at two primary parts of the Vision Transformer: first, to replace the existing position-wise linear projection for the attention operation with convolutional projection, and second, to use hierarchical multi-stage structure to enable varied resolution of 2D reshaped token maps. Whereas, in our work, a convolutional feature map to low dimensional space (Posit8) is given as input to the transformer. We have three convolutional stacks at the input to the transformer. Then comes the transformer block, which contains Layer Normalisation(LN), Multi-Head Self Attention (MHSA), and the Multi-Layer Perceptron (MLP). The output of the transformer is given to the MLP head as in the original vision transformer.

The feature map in layer l denoted Y_l is computed as

$$Y_l = CONV(\text{Posit Encoded}(Y_{l-1}) * \text{Posit Encoded}(K_l)) , L = 1 \ldots l \quad (9)$$

$$Y_l = MHSA(LN(Y_{l-1})) + Y_{l-1} , L = 1 \ldots l \quad (10)$$

$$Y_l = MLP(LN(Y_l)) + Y_l , L = 1 \ldots l \quad (11)$$

$$Z = LN(Y_l^0) \quad (12)$$

The step-wise equations for our stacked vision transformer is shown in Eq. 9, 10, 11, and 12. Where CONV is Convolution, MHSA is Multi-Head Self-Attention, MLP is Multi-Layer Perceptron and LN is Layer Normalization.

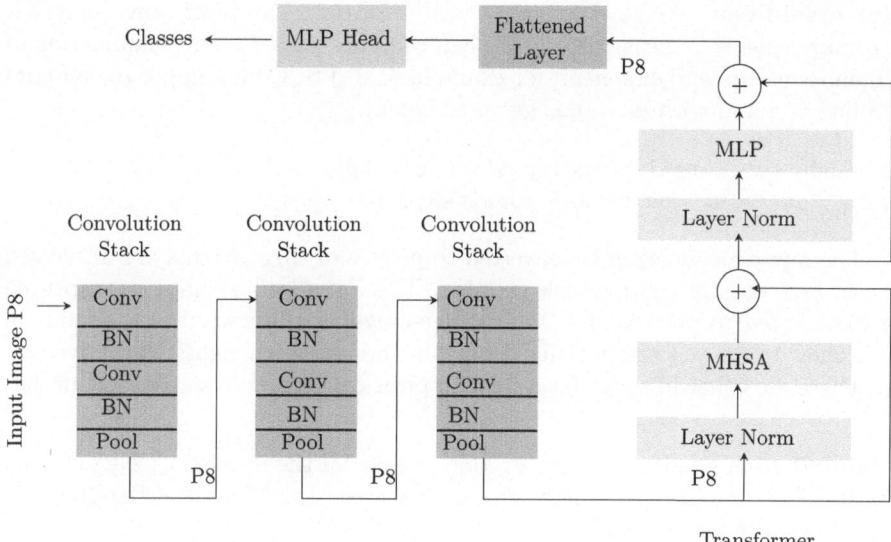

Fig. 10. Stacked Vision Transformer (Stacked ViT)

The ViT receives input as a feature map of the previous max-pooling layer. Stacked ViT improves Vision Transformer (ViT) in performance and efficiency by introducing convolutions into ViT to yield the best of both designs. These changes introduce desirable properties of convolutional neural networks to the ViT architecture (i.e. shift, scale, and distortion invariance) while maintaining the merits of Transformers (i.e. dynamic attention, global context, and better generalization) [15].

In our implementation, we have utilized the Universal Number Library [18] to perform the conversion of floating-point values to the Posit Format with a configuration of $< 8, 0 >$. The input image array is initially converted to posit representation, and during the first pass of network training, the weights are initialized. We capture these initialized weights and convert them to Posit values as well. The flow of posit values within the network can be observed in Fig. 10, where the symbol $P8$ signifies the utilization of posit values.

5   Implementation

Conversion to Float32 and Bfloat16. To convert a number into Float32, we have built-in command using numpy library and that is: numpy.float32(number).

To convert a given number to Bfloat16, we can type cast the given number to Bfloat16 using TensorFlow: tf.cast(number, tf.bfloat16). Likewise, to convert a number to posit, we are making use of a library called Universal [18].

Universal Posit. We have used Universal[1] library [18] for posit conversion as it is comprehensive and easy to use. We can compute posit for any combination of a number of bits and exponent, for example, $< 8, 0 >$. Code snippet to compute a n-bits = 8 and with es = 0, is given as below:

```
#include <universal/number/posit/posit.hpp>
using Real = sw::universal::posit<8,0>;
```

The input image array is converted to posit and then weights are initialized in the first pass of the network training. The initialized weights are captured and converted to posit values. We use get-weight and set-weight commands in tensorflow to extract the initialized weights and set new weights respectively.

There are other libraries for a float to posit conversion example, Soft posit[2].

Training with Posit. Posit integration is done for the Stacked CNN, ViT, and Stacked ViT. We have considered three databases here: Cifar-10, Cifar-100, and Tiny Imagenet for our study.

System Specifications: processor = 11th Gen Intel® Core™ i3-1115G4 @ 3.00 GHz × 4, Memory = 7.6 GB, OS name = Ubuntu 20.04.3 LTS, OS Type = 64-bit.

Stacked CNN specifications: filter_Size = 32,64,128, kernel_size = 3X3, strides = 1X1, max_pool = 2X2, batch_size = 32, learning_rate = 0.001, loss = Sparse_Categorical_Crossentropy, Optimizer = Adam.

ViT specifications: learning_rate = 0.001, weight_decay = 0.0001, batch_size = 512, image_size = 32, patch_size = 6, num_patches = (image_size patch_size) ** 2, projection_dim = 32, num_heads = 4, transformer_units = [projection_dim * 2, projection_dim], transformer_layers = 8, mlp_head_units = [2048, 1024], loss = Sparse_Categorical_Crossentropy, Optimizer = AdamW.

Stacked ViT specifications: filter_size = 32, kernel_size = 3X3, learning_rate = 0.001, batch_size = 500, transformer_layers = 8, mlp_head_units = [2048, 1024], loss = Sparse_Categorical_Crossentropy, Optimizer = AdamW.

Dataset. We are using Cifar-10, Cifar-100, and Tiny Imagenet in our experiments.

Cifar-10: The Cifar-10 dataset is a popular dataset used in machine learning and computer vision for image classification tasks. The classes include common objects and animals such as "airplanes", "automobiles", "birds", "cats", "deer", "dogs", "frogs", "horses", "ships", and "trucks". Cifar-10 contains a collection

---

[1] https://github.com/stillwater-sc/universal.
[2] https://posithub.org/docs/PositTutorial_Part1.html.

of 60,000 small, 32×32 pixel color images, which are divided into 10 classes, with each class having 6,000 images. The training set consists of 50,000 images (5,000 images per class). The testing set contains 10,000 images (1,000 images per class). All images are in color, with each pixel having three color channels (RGB: Red, Green, and Blue).

Cifar-100: The Cifar-100 dataset is an extension of the Cifar-10 dataset, designed for more fine-grained image classification tasks. It contains a larger collection of 60,000 color images, with 32×32 pixels, like Cifar-10, but these images are divided into 100 classes, with each class having 600 images. The dataset includes 100 classes, each representing a fine-grained category of objects or animals. Examples of classes include "beaver," "dolphin," "television," "maple_tree," and "shark". The training set consists of 50,000 images (5,000 images per class). The testing set contains 10,000 images (1,000 images per class). All images are in color, with each pixel having three color channels (RGB: Red, Green, and Blue).

Tiny Imagenet: The Tiny ImageNet dataset is a smaller version of the popular ImageNet dataset and is often used for image classification tasks and benchmarking in the field of computer vision. The Tiny ImageNet dataset consists of 200 classes, each representing a different object or category. All images in the dataset are scaled down to a small size of 64×64 pixels. The dataset contains a total of around 120,000 images. The training set contains 100,000 images and the test set of around 20,000 images is used to evaluate the model's performance.

6 Results and Discussion

Posit integration is done for the Stacked CNN, Vision Transformer, and Stacked ViT. The number of parameters used in Stacked CNN, ViT and Stacked ViT is shown in Table 3. Stacked ViT has a comparably less number of parameters than Stacked CNN and ViT. Networks with fewer parameters require less memory and computational resources during training and inference. This can be advantageous for deploying models on resource-constrained devices or for speeding up training.

Table 4 shows the training accuracy of Stacked CNN, ViT, and Stacked ViT with two datasets, Cifar-10 and Cifar-100, and with three datatypes (Posit8, Bfloat16, and Float32). The results are compared with the original ViT [3] and CvT [15]. The main purpose of using Posit is to take advantage of its low bit representation while at the same time having comparable performance with its peers such as Bfloat16 and IEEE floating point representation (Float32). Our Stacked ViT has shown better training accuracy when compared to Stacked CNN and ViT. Especially, with Cifar-100, our Stacked ViT has outscored the traditional Stacked CNN and ViT.

The time taken to complete a hundred epochs using Stacked CNN posit $< 8, 0 >$, ViT posit $< 8, 0 >$ and Stacked ViT posit $< 8, 0 >$ is shown in Table 5. The time taken by Stacked ViT posit $< 8, 0 >$ is outstanding when compared to the time taken by Stacked CNN posit $< 8, 0 >$ and ViT posit $< 8, 0 >$.

Stacked ViT's Training loss versus the number of epochs and Validation loss versus the number of epochs is shown in Fig. 11. The curve shows that there is no significant validation and training loss in spite of using low-bit posit encoded ViT.

The classification accuracy and training time of Tiny Imagenet with Stacked CNN, ViT, and Stacked ViT is shown in Table 6. Stacked CNN creates a competitive baseline and Stacked ViT has matched the Stacked CNN in performance. It is worth noting that the time taken by Stacked ViT with Posit8 and Float32 (336 min and 345 min respectively) is better than the Stacked CNN with Posit8 and Float32 (588 min and 592 min) baseline making low-precision Stacked ViT a better alternative for ViT and Stacked CNN.

Although the IEEE 754 standard had long defined a 16-bit float representation, several other alternatives targeting mixed-precision training have also emerged. However, their varying numerical properties and differing hardware characteristics among other things make them more or less suitable for the task.

Table 3. Total trainable parameters for Stacked CNN, ViT and Stacked ViT

	Stacked CNN	ViT	Stacked ViT
Cifar-10	2,396,330	3,922,122	3,227,754
Cifar-100	2,488,580	4,014,372	3,320,004
Tiny Imagenet	8,882,536	9,034,472	6,565,232

Table 4. Stacked CNN, ViT and Stacked ViT with accuracy for 100 epochs (using Cifar-10 and Cifar-100), NA: Not Available

	Cifar-10		Cifar-100	
Number System	Training(%)	Validation (%)	Training(%)	Validation (%)
Stacked CNN posit $< 8, 0 >$	97.66	82.32	94.84	51.33
Stacked CNN Bfloat16	97.63	82.29	94.81	51.03
Stacked CNN Float32	97.67	82.69	94.76	52.40
ViT posit $< 8, 0 >$	95.60	70.94	91.77	42.28
ViT Bfloat16	95.81	71.84	91.82	42.50
ViT Float32	95.50	70.77	91.57	39.50
Stacked ViT posit $< 8, 0 >$	**98.44**	79.33	**96.17**	46.93
Stacked ViT Bfloat16	**98.57**	70.56	**95.98**	46.42
Stacked ViT Float32	**98.51**	79.23	**96.07**	47.83
ViT-B/16 [15]	98.95	NA	91.67	NA
CvT-13 [15]	98.83	NA	91.11	NA

Table 5. Time taken to train 100 epochs in secs for posit $< 8, 0 >$ of Stacked CNN, ViT, Stacked ViT

	Stacked CNN (sec)	ViT (sec)	Stacked ViT (sec)
Cifar-10	8796	7346	**4676**
Cifar-100	8374	6880	**4513**

Table 6. Stacked CNN, ViT and Stacked ViT with accuracy and time taken for 56 epochs using Tiny Imagenet

	Tiny Imagenet		
Number System	Training (%)	Validation (%)	Training Time (minutes)
Stacked CNN posit $< 8, 0 >$	94.85	30.25	590
Stacked CNN Float32	94.48	29.67	592
ViT posit $< 8, 0 >$	61.82	24.34	703
ViT Float32	61.27	24.80	704
Stacked ViT posit $< 8, 0 >$	**93.74**	29.39	**336**
Stacked ViT Float32	**93.76**	28.14	**340**

In Table 7, we have a comparison of the performance of three different models: Stacked CNN, ViT, and Stacked ViT. For each model, we have recorded the best accuracy achieved (out of three runs: Float32, Bfloat16, and Posit8) and the time taken to train the model. Stacked ViT performs better than Stacked CNN and ViT with Cifar-10 and Cifar-100. Stacked ViT takes less time to train when compared to Stacked CNN and ViT and is indicated by the down arrow.

Table 7. Stacked CNN, ViT and Stacked ViT with (best out of three datatypes: Float32, Bfloat16 and Posit8) accuracy and time taken to train the model

	Stacked CNN		ViT		Stacked ViT	
	Accuracy	Time	Accuracy	Time	Accuracy	Time
Cifar-10	97.66	↑	95.81	↑	**98.57**	↓
Cifar-100	94.84	↑	91.82	↑	**96.17**	↓
TinyImagenet	**94.85**	↑	61.82	↑	93.76	↓

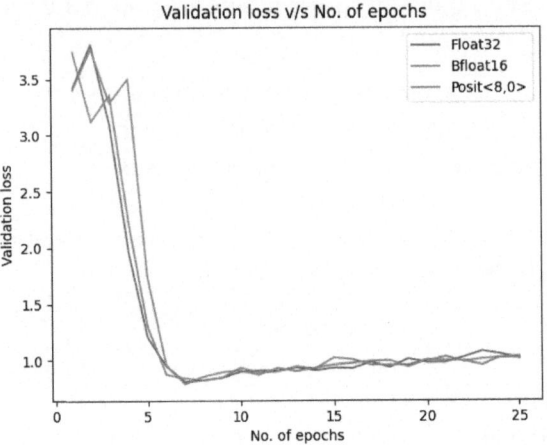

(a) Stacked ViT with validation loss (using Cifar-10)

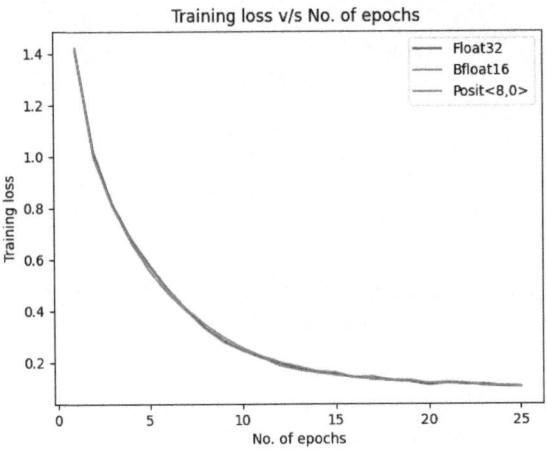

(b) Stacked ViT with training loss (using Cifar-10)

Fig. 11. Training and Validation loss with Stacked ViT

7   Conclusion

In conclusion, our study unequivocally demonstrates the remarkable benefits of employing low-bit Posit Arithmetic in Vision Transformers. Our approach offers higher accuracy, reduced training time, decreased storage requirements, and lower power consumption compared to traditional methods. By harnessing the unique advantages of posit number representation, our low-bit Stacked ViT model not only outperforms the conventional Stacked CNN model but also achieves comparable or superior accuracy compared to state-of-the-art ViT models, all while significantly reducing the number of parameters, floating-point operations (FLOPs), and training time.

These results showcase the potential of posit-enabled low-precision Vision Transformers as a promising solution for AI applications operating within strict computational constraints. Moreover, the versatility of low-bit posits extends to various neural network architectures, opening doors for future research in the fields of computer vision and machine learning. Our findings contribute to the development of more efficient and accurate models for AI, meeting the escalating demand for high-performance, energy-efficient computing.

Furthermore, our utilization of Posit Arithmetic underscores the transformative potential of this new number system in revolutionizing the design and implementation of neural networks in future applications. Our ongoing work will delve into optimizing the allocation of bits within the posit format for different layers or components of the neural network, further enhancing the efficiency and adaptability of our approach.

References

1. Gustafson, J.L., Yonemoto, I.: Beating floating point at its own game: posit arithmetic. Supercomput. Front. Innov. **4**(2), 71–86 (2017)
2. Ho, N.M., Nguyen, D.T., Silva, H.D., Gustafson, J.L., Wong, W.F., Chang, I.J.: Posit arithmetic for the training and deployment of generative adversarial networks. In: 2021 Design, Automation and Test in Europe Conference and Exhibition (DATE), p. 13501355. IEEE (2021)
3. Dosovitskiy, A., et al.: An image is worth 16×16 words: transformers for image recognition at scale. arXiv preprint arXiv:2010.11929 (2020)
4. Burgess, N., Milanovic, J., Stephens, N., Monachopoulos, K., Mansell, D.: Bfloat16 processing for neural networks. In: 2019 IEEE 26th Symposium on Computer Arithmetic (ARITH), pp. 88–91 (2019). https://doi.org/10.1109/ARITH.2019.00022
5. Raposo, G., Tomás, P., Roma, N.: PositNN: training deep neural networks with mixed low-precision posit. In: 2021 IEEE International Conference on Acoustics, Speech and Signal Processing (ICASSP), ICASSP 2021, pp. 7908–7912 (2021). https://doi.org/10.1109/ICASSP39728.2021.9413919
6. Murillo, R., Del Barrio, A.A., Botella, G.: Deep PeNSieve: a deep learning framework based on the posit number system. Digit. Sig. Process. **102**, 102762 (2020)
7. Carmichael, Z., Langroudi, H.F., Khazanov, C., Lillie, J., Gustafson, J.L., Kudithipudi, D.: Deep positron: a deep neural network using the posit number system. In: 2019 Design, Automation and Test in Europe Conference and Exhibition (DATE), pp. 1421–1426. IEEE (2019)
8. Lu, J., Fang, C., Mingyang, X., Lin, J., Wang, Z.: Evaluations on deep neural networks training using posit number system. IEEE Trans. Comput. **70**(2), 174–187 (2020)
9. De Dinechin, F., Forget, L., Muller, J.M., Uguen, Y.: Posits: the good, the bad and the ugly. In: Proceedings of the Conference for Next Generation Arithmetic 2019, pp. 1–10 (2019)

10. Ho, NM., De Silva, H., Gustafson, J.L., Wong, W.F.: Qtorch+: next generation arithmetic for Pytorch machine learning. In: Gustafson, J., Dimitrov, V. (eds.) Next Generation Arithmetic, CoNGA 2022. LNCS, vol. 13253, pp. 31–49. Springer, Cham (2022). https://doi.org/10.1007/978-3-031-09779-9_3
11. Posithub.org: Posit Standard Documentation Release 3.2-draft (2018). https://posithub.org/docs/posit_standard.pdf. Accessed 03 Jan 2022
12. Vaswani, A., et al.: Attention is all you need. In: Advances in Neural Information Processing Systems, vol. 30 (2017)
13. Lee, S.H., Lee, S., Song, B.C: Vision transformer for small-size datasets. arXiv preprint arXiv:2112.13492 (2021)
14. Gani, H., Naseer, M., Yaqub, M.: How to train vision transformer on small-scale datasets? arXiv preprint arXiv:2210.07240 (2022)
15. Wu, H., et al.: CvT: introducing convolutions to vision transformers. In: Proceedings of the IEEE/CVF International Conference on Computer Vision, pp. 22–31 (2021)
16. Ioffe, S., Szegedy, C.: Batch Normalization: accelerating deep network training by reducing internal covariate shift. In: International Conference on Machine Learning, pp. 448–456. PMLR (2015)
17. "IEEE Standard for Floating-Point Arithmetic" in IEEE STD 754-2019 (Revision of IEEE 754-2008), pp. 1–84, 22 July 2019. https://doi.org/10.1109/IEEESTD.2019.8766229
18. Omtzigt, E., Theodore, L., Gottschling, P., Seligman, M., Zorn, W.: Universal numbers library: design and implementation of a high-performance reproducible number systems library. arXiv preprint arXiv:2012.11011 (2020)
19. Posit working group, "Posit Standard Documentation, Release 3.2-draft," (2018). https://posithub.org/docs/posit standard.pdf. Accessed 24 Sep 2020
20. Gustafson, John. "Posit Arithmetic. mathematica Notebook describing the posit number system (2017)
21. Ghosh, A., Sufian, A., Sultana, F., Chakrabarti, A., De, D.: Fundamental concepts of convolutional neural network. In: Balas, V.E., Kumar, R., Srivastava, R. (eds.) Recent Trends and Advances in Artificial Intelligence and Internet of Things. ISRL, vol. 172, pp. 519–567. Springer, Cham (2020). https://doi.org/10.1007/978-3-030-32644-9_36
22. Buoncristiani, N., Shah, S., Donofrio, D., Shalf, J.: Evaluating the numerical stability of posit arithmetic. In: 2020 IEEE International Parallel and Distributed Processing Symposium (IPDPS), pp. 612–621. IEEE (2020)
23. Zhou, D., et al.: Deepvit: towards deeper vision transformer. arXiv preprint arXiv:2103.11886 (2021)
24. Guo, J., et al: CMT: convolutional neural networks meet vision transformers. In: Proceedings of the IEEE/CVF Conference on Computer Vision and Pattern Recognition, pp. 12175–12185 (2022)
25. He, K., Zhang, X., Ren, S., Sun, J.: Deep residual learning for image recognition. In: Proceedings of the IEEE Conference on Computer Vision and Pattern Recognition, pp. 770–778 (2016)
26. Liu, Z., et al .: Swin transformer: hierarchical vision transformer using shifted windows. In: Proceedings of the IEEE/CVF International Conference on Computer Vision, pp. 10012–10022. (2021)
27. Touvron, H., Cord, M., Douze, M., Massa, F., Sablayrolles, A., Jégou, H.: Training data-efficient image transformers and distillation through attention. In: International Conference on Machine Learning, pp. 10347–10357. PMLR (2021)

28. Cococcioni, M., Rossi, F., Ruffaldi, E., Saponara, S.: Fast approximations of activation functions in deep neural networks when using posit arithmetic. Sensors **20**(5), 1515 (2020)
29. Gustafson, J.L.: The End of ERROR: Unum Computing. CRC Press (2017)
30. Lu, J., et al.: Training deep neural networks using posit number system. In: 2019 32nd IEEE International System-on-Chip Conference (SOCC), pp. 62–67. IEEE (2019)

Iterative Refinement with Low-Precision Posit Arithmetic

James Quinlan[1]([✉])[iD] and E. Theodore L. Omtzigt[2][iD]

[1] University of Southern Maine, Portland, ME 04104, USA
james.quinlan@maine.edu
[2] Stillwater Supercomputing Inc., El Dorado Hills, CA 95762, USA
tomtzigt@stillwater-sc.com

Abstract. This study examines the mixed-precision iterative refinement technique using posit numbers instead of standard IEEE floating-point. The process is applied to a general linear system $Ax = b$ where A is a large sparse matrix. Multiple scaling strategies, including row and column equilibration, scale matrix entries into higher-density regions of machine numbers before performing the $O(n^3)$ factorization operation. Low-precision LU factorization followed by forward/backward substitution yields an initial estimate. The residual $r = b - Ax$ is computed to a higher precision with a deferred rounding mechanism, then used as the right-hand side in a new linear system $Ac = r$. The corrector c is calculated and used to refine the previous solution. Results show a 16-bit posit configuration coupled with equilibration yields accuracy comparable to IEEE half-precision (fp16), showing potential for balancing efficiency and accuracy.

Keywords: mixed-precision · iterative refinement · *Universal* Library

1 Introduction

Next-generation arithmetic is increasingly expected to address issues with power and performance in computing. Both high-performance computing (HPC) and artificial intelligence (AI) fields are exploring nontraditional floating-point representations and arithmetic [24,33,38,47]. Low precision formats such as IEEE's half-precision (fp16) and Google's brain float (`bfloat16`) can run at least twice the speed [17] and require only a fraction of the storage [49]. Such features, desirable for deep learning models [22,47], are also crucial in model simulation [45]. Manufacturers, including Intel [32] and NVIDIA [33], are customizing hardware to support these alternate formats. Recently, hardware for the posit number system [23] has also been introduced [42,46].

Our focus in this paper is to examine a low precision format in approximating the solution of a general linear system of equations $Ax = b$ where $A \in \mathbb{R}^{n \times n}$ is a nonsingular matrix and $b \in \mathbb{R}^n$. Such systems are encountered in many scientific and industrial applications [37] and one of the most frequently occurring problems in computing [18, p.16].

© The Author(s), under exclusive license to Springer Nature Switzerland AG 2024
M. Michalewicz et al. (Eds.): CoNGA 2024, LNCS 14666, pp. 74–90, 2024.
https://doi.org/10.1007/978-3-031-72709-2_3

In this work, we represent the entries of A using a low-precision posit format to leverage the benefits of reduced power consumption and improved performance. However, we strive to maintain comparable levels of backward and forward errors, as accurate solutions are crucial in many applications spanning various domains, such as mathematical physics [4], supercomputing [27], optimization [40], and molecular biology [41].

The paper is organized as follows: Sect. 2 presents iterative refinement, mixed-precision iterative refinement, and some previous work on mixed-precision iterative refinement in the literature. Section 3 covers the posit number system, gives an example, provides a table of properties, and compares it with other formats. The details of our numerical experiments are presented in Sect. 4, including the matrix test-suite, conversion algorithms to low-precision posits, and technical specifications for replication. Results are presented in Sect. 5, and a summary is presented in Sect. 6. We take $\|\cdot\| = \|\cdot\|_\infty$ to be the infinity norm for its practical error estimation [19, p. 123-124].

2   Iterative Refinement

Iterative refinement (Algorithm 1), introduced by Wilkinson [48], consists of three steps to improve an approximate solution of a system of equations $Ax = b$ when A is ill-conditioned. First, an initial solution x_0 is computed. This step uses Gaussian elimination (i.e., LU), which requires $O(n^3)$ operations. After an initial approximation, x_0, the method computes the residual $r_0 = b - Ax_0$ validates the intial solution's accuracy. A correction is then computed by solving $Ac_0 = r_0$ and the approximate is updated by $x_1 = x_0 + c_0$. The last two steps are iterated until convergence. For a detailed discussion of stopping criteria, refer to [5,15]. However, when A is extremely ill-conditioned, iterative refinement may not lead to convergence, and additional modifications are needed.

Algorithm 1. Iterative refinement

Input: $A \in \mathbb{R}^{n \times n}$, $b \in \mathbb{R}^n$, and maximum number of iterations N
Output: A vector approximating solution x to $Ax = b$
1: Solve $Ax_0 = b$
2: **for** $i = 0 : N$ **do**
3:     $r_i = b - Ax_i$
4:     Solve $Ac_i = r_i$
5:     $x_{i+1} \leftarrow x_i + c_i$
6:     **if** converged **then**
7:         return x_{i+1}
8:     **end if**
9: **end for**

Mixed-precision iterative refinement (Algorithm 2) is considered one of the most promising techniques for obtaining a high-precision solution to a linear equation while leveraging low-precision arithmetic for expensive computations. It attempts to balance computational performance and solution accuracy. The central idea behind mixed-precision iterative refinement is to perform the $O(n^3)$ operations in Gaussian elimination using low-precision arithmetic, storing the results in working-precision. Subsequently, the approximation is iteratively refined by solving the correction equation using the high-precision residual, as outlined in the algorithm. In Step 6, we use the *quire* memory register to defer rounding in our experiments. This iterative process allows the solution to reach the desired level of accuracy while minimizing the computational overhead associated with high-precision operations.

Several researchers have investigated mixed-precision iterative refinement. Langou et al. introduced a mixed-precision iterative refinement approach where the matrix factorization step is performed in fp32, while the remaining operations are carried out in fp64 [3,36]. Carson and Higham have recently explored versions of mixed-precision iterative refinement utilizing three precisions [9,10]. Haidar et al. have developed parallel implementations of these mixed-precision iterative refinement techniques, as described in [26]. Higham et al. [30] developed a mixed-precision GMRES-based iterative refinement algorithm using two-sided scaling to convert to half-precision format, which moves the elements of the largest magnitude close to the overflow threshold. In [7], researchers applied iterative refinement to symmetric matrices using various posit configurations (see Sect. 3). Iterative refinement remains an active area of research, as evidenced by various contributions [1,3,9,10,24,28].

Algorithm 2. Mixed-precision iterative refinement

Input: $A \in \mathbb{R}^{n \times n}$ and $b \in \mathbb{R}^n$
Output: A vector $\hat{x}$ approximating solution x to $Ax = b$
 1: Factor $A = LU$ in low-precision
 2: Solve $LUx_0 = b$ in working-precision
 3: $i \leftarrow 0$
 4: $r_0 = b - Ax_0$
 5: **while** $||r_i|| > \epsilon(\|A\|_\infty \|x_i\|_\infty + \|b\|_\infty)$ **do**
 6:     $r_i = b - Ax_i$ compute in high-preicsion
 7:     Solve $Ac_i = r_i$
 8:     $x_{i+1} \leftarrow x_i + c_i$
 9:     $i \leftarrow i + 1$
10: **end while**

---

3   Posit Number System

The *Posit Number System* (PNS) is a set of machine numbers, denoted by $\mathbb{P}$, representing the real numbers $\mathbb{R}$ in binary developed by [23] in 2017. Two parameters characterize a posit representation: (i) the total number of bits n and (ii) the maximum number of exponent bits E. A *standard* n-bit posit ($E = 2$) [21], is decoded as:

$$x = ((1 - 3s) + f) \times 2^{(1-2s)(4k+e+s)} \tag{1}$$

where the sign bit s is 0 if $x \geq 0$ and 1 for $x < 0$. The exponent is $0 \leq e \leq E$. In the standard, the exponent $e \in \{0, 1, 2, 3\}$. For positive posits, (1) simplifies to,

$$x = (1 + f) \times 2^{4k+e}$$

The most notable distinction from IEEE-754 is the posit's dynamic *regime*[1] field. While the exponent is a power-of-two scaling determined by the exponent bits, the regime is a power-of-16 scaling determined by (2) where r is the number of identical bits in the run before an opposite terminating bit, r^c.

$$k = \begin{cases} -r & \text{if run of 0's length } r \\ r - 1 & \text{if run of 1's length } r, \end{cases} \tag{2}$$

and the normalized fraction $f \in [0, 1)$ given by

$$f = 2^{-m} \sum_{i=0}^{m-1} f_i 2^i$$

with m the number of fraction bits ranging between 0 and $\max(0, n-5)$ and $f_i \in \{0, 1\}$ is the ith bit in fraction bitstring. The standard n-bit posit configuration is denoted as posit$\langle$n,2$\rangle$. Often the "2" is dropped, for example, the standard 64-bit posit, posit$\langle$64,2$\rangle$, would be written as posit64.

For example, Fig. 1 shows the bitstring of the number 3.5465×10^{-6} encoded as a posit$\langle$16,2$\rangle$. In this example, we have a run length of $r = 5$, therefore $k = -5$. The exponent $e = 1$ and $m = 7$ fraction bits.

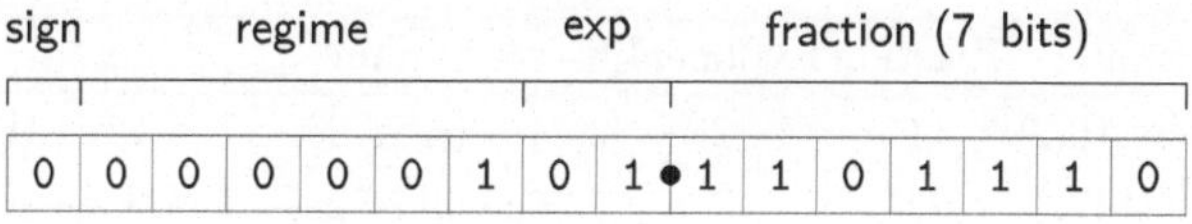

sign	regime	exp	fraction (7 bits)

| 0 | 0 | 0 | 0 | 0 | 0 | 1 | 0 | 1 | 1 | 1 | 0 | 1 | 1 | 1 | 0 |

Fig. 1. A posit$\langle 16, 2 \rangle$ with 7-bit fraction and 2-bit exponent.

The decoding given by (1) is:

$$\left((1 - 3 \cdot 0) + \frac{1}{2} + \frac{1}{4} + \frac{0}{8} + \frac{1}{16} + \frac{1}{32} + \frac{1}{64} + \frac{0}{128} \right) \times 2^{4(-5)+1}. \tag{3}$$

A posit visualization tool is available online [34].

The posit standard specifies the use of an accumulation register called a *quire* to apply the *Karlsruhe Accurate Arithmetic Approach* (KAAA) [35] to

---

[1] There is always at least one regime bit, and for $n > 2$, there are at least two bits. For $n > 2$, $1 \leq r \leq n - 1$, therefore $-(n - 1) \leq k \leq n - 2$.

defer rounding. According to the Standard [21, p. 6], "a quire value is either NaN or an integer multiple of the square of the minimum positive value, represented as a 2's complement binary number with $16n$ bits." For posit$\langle 16,2 \rangle$, there are 256 bits reserved for this accumulator. A single round is performed to produce the final representation.

Table 1 lists properties, including the maximum consecutive integer, the range of fraction lengths, quire precision, and the minimum and maximum positive representable number in the posit standard. There are always $2^{n-2} + 1$ values in the interval $[0, 1]$, regardless of the number of exponents bits. Table 2 displays comparisons between various posit configurations and floating-point formats.

Table 1. Posit Properties under the standard $E = 2$.

Property	Value
Fraction length	0 to $\max(0, n - 5)$ bits
Minimum Positive Value	2^{-4n+8}
Maximum Positive Value	2^{4n-8}
Maximum Consecutive Integer	$\lceil 2^{\lfloor 4(n-3)/5 \rfloor} \rceil$
Quire format precision	$16n$ bits
Quire sum limit	2^{23+4n}
Number of values in $[0, 1]$	$2^{n-2} + 1$

Table 2. Specifications of multiple binary formats for comparison including IEEE (half, single, and double precision), Google's bfloat16, and three standard posit configurations. The precision of the arithmetic is measured by the unit round-off, listed below as the distance from 1.0 to the next larger representable value.

Format	Fraction	Exponent	u	min	max
posit$\langle 16,2 \rangle$	11*	2	$2.44e{-}04$	$1.39e{-}17$	$7.21e{+}16$
posit$\langle 32,2 \rangle$	27*	2	$7.45e{-}09$	$7.52e{-}37$	$1.33e{+}36$
posit$\langle 64,2 \rangle$	59*	2	$1.73e{-}18$	$2.21e{-}75$	$4.52e{+}74$
bfloat16	8	8	$3.91e{-}03$	$1.18e{-}38$	$3.39e{+}38$
fp16	11	5	$4.88e{-}04$	$6.10e{-}05$	$6.55e{+}04$
fp32	24	8	$1.19e{-}07$	$1.18e{-}38$	$3.40e{+}38$
fp64	53	11	$1.11e{-}16$	$2.22e{-}308$	$1.80e{+}308$

* denotes variable fraction bits (maximum number of bits listed)

4   Numerical Experiments

We perform numerical experiments to compare the convergence of the solution to the system $Ax = b$ using mixed-precision iterative refinement (Algorithm

2) with the posit number system. Three algorithms converting to low-precision posits to perform the $O(n^3)$ operations will be examined. For all systems, the exact solution is taken to be $x = (1, 1, \ldots, 1)^T$.

Posit Configurations. The incorporation of the quire in our experiments facilitates the utilization of two posit configurations: posit$\langle 32,2 \rangle$ for maintaining working precision and posit$\langle 16,2 \rangle$ for conducting low-precision calculations. This approach contrasts with the requisite use of three precisions in analogous experiments performed within the IEEE floating-point framework [24–26, 29, 30].

Test-Suite. Table 3 lists ten nonsingular square matrices extracted from the *SuiteSparse Matrix Collection*[2] [13] sorted by the unique identifier (ID), along with key characteristics such as the size, percent of nonzero entries, maximum and minimum absolute (nonzero) entry, and condition number. We restrict the size the matrices to $n \leq 240$ due time required for LU factorization without native hardware support for posits. Figure 2 displays the distribution of unique unsigned matrix elements with a y-axis of the relative frequency on a logarithmic scale. These matrices represent problems in diverse fields such as computational fluid dynamics (CFD), chemical simulation, materials science, optimal control, structural mechanics, and 2D/3D sequencing. The matrices were chosen to facilitate a (near) comparative analysis with Higham et al. [30], (and others), who employed fp16 for the low-precision computations, fp32 as the working precision, and fp64 for residual calculations in the iterative refinement process. However, one difference is nearly all the entries lie within the dynamic range of posit16. Figure 2 shows the distribution of unsigned nonzero matrix entries on a logarithmic scale.

Table 3. Selected matrices representing various disciplines from the Suite Sparse Matrix Collection. ID is the matrix identifier as listed in the collection. A '*' denotes a symmetric matrix.

ID	Matrix	n	nnz(%)	max $|a_{ij}|$	min $|a_{ij}| \neq 0$	$\kappa_\infty(A)$
6	arc130	130	6.14	$1.05e{+}05$	$7.71e{-}31$	$1.20e{+}12$
23*	bcsstk01	48	17.36	$2.47e{+}09$	$3.33e{+}03$	$1.60e{+}06$
27*	bcsstk05	153	10.35	$1.43e{+}04$	$4.65e{-}10$	$3.53e{+}04$
206*	lund_a	147	11.33	$1.50e{+}08$	$1.22e{-}04$	$5.44e{+}06$
217*	nos1	237	1.81	$1.22e{+}09$	$8.00e{+}05$	$2.53e{+}07$
232	pores_1	30	20.00	$2.46e{+}07$	$4.00e{+}00$	$2.49e{+}06$
239	saylr1	238	1.99	$3.06e{+}08$	$7.19e{-}04$	$1.59e{+}09$
251	steam1	240	3.90	$2.17e{+}07$	$1.48e{-}07$	$3.11e{+}07$
263	west0132	132	2.37	$3.16e{+}05$	$3.31e{-}05$	$1.05e{+}12$
298	bwm200	200	1.99	$6.15e{+}02$	$4.00e{+}00$	$2.93e{+}03$

---

[2] Formerly known as the University of Florida Sparse Matrix Collection.

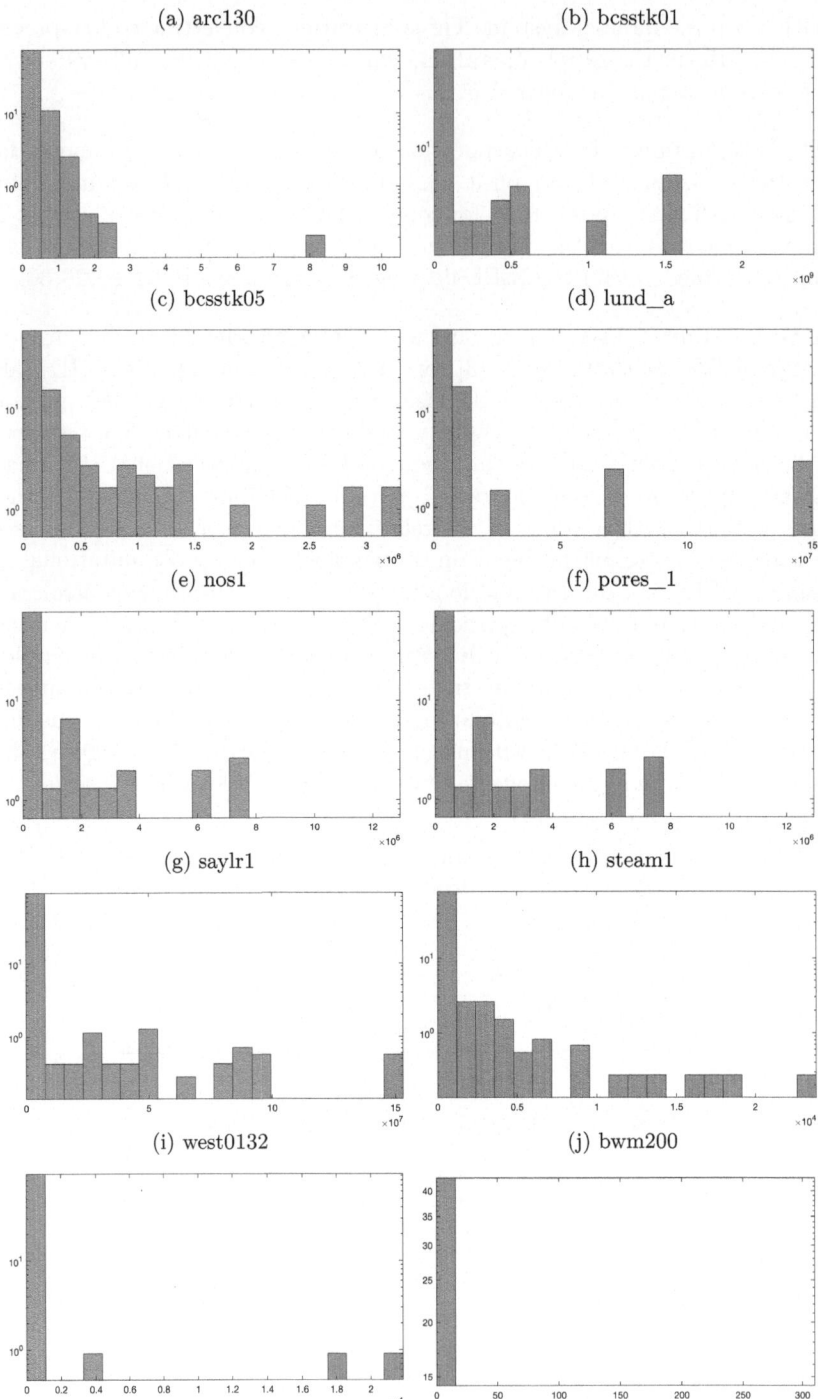

Fig. 2. Distribution of unique unsigned matrix elements with y-axis set to log(percent frequency). The largest portion of entries are contained in the first bin.

Converting to Low Precision Posits. Many matrices from computational sciences will have entries outside the dynamic range of these low-precision configurations. In IEEE half (or single) precision, the elements of A may overflow or under-flow during the transition to lower precision. However, posits do not overflow or underflow, and matrix entries exceeding the dynamic range will be mapped to the maximum or minimum signed value. Algorithm 3 converts a higher precision posit to a lower precision.

Algorithm 3. Round and replace (*Default*: Cast to low-precision posit)

Input: $A \in \mathbb{R}^{n \times n}$
Output: $A^{(L)}$ lower configuration
1: $A^{(L)} = \text{fl(A)}$ using posit$\langle<,1\rangle 6,2>$
2: $\forall i,j$ such that $|a_{ij}^{(L)}| > x_{\max}$, set $a_{ij}^{(L)} = \text{sgn}(a_{ij})\, x_{\max}$
3: $\forall i,j$ such that $|a_{ij}^{(L)}| < x_{\min}$, set $a_{ij}^{(L)} = \text{sgn}(a_{ij})\, x_{\min}$

The second approach, Algorithm 4, follows the method proposed in [30]. In this approach, the matrix is scaled before rounding. After scaling and rounding the matrix, Algorithm 2 will be applied. Note that the linear system's right-hand side vector b must also be scaled by the same factor μ to preserve the equality of the system after scaling the matrix. This scaled right-hand side is denoted as μb.

Algorithm 4. Scale matrix entries, then round to low-precision

---

Input: $A \in \mathbb{R}^{n \times n}$
Output: $A^{(L)}$ low-precision
1: $A = \mu A$ for $\mu \in \mathbb{R}$
2: $A^{(L)} = \text{fl}(A)$

The third conversion strategy uses two-sided diagonal scaling. Since the con-dition number $\kappa(A) = ||A|| \cdot ||A^{-1}||$ affects error bounds, for $\kappa(A) \gg 1$, iter-ative refinement may never converge [9]. To mitigate this issue, the system is often rescaled to reduce the condition number, aiming to improve convergence. However, it is crucial to note that scaling strategies can be unreliable, as they may produce more ill-conditioned matrices [18]. Therefore, it is generally recom-mended to apply scaling strategies on a problem-by-problem basis, taking into consideration the significance of the entries a_{ij}, the units of measurement, and error in the data [20].

The primary technique used in this work is to apply simple two-sided row and column scalings to the matrix A. These scalings aim to equilibrate A, as described in the references [18] and [20] to reduce the condition number of A.

In particular, equilibration works to solve $Ax = b$ by first mutating the system by premultiplying by the diagonal matrix D_1^{-1} to get $D_1^{-1}Ax = D_1^{-1}b$. The goal is to choose D_1^{-1} so that $D_1 A$ has approximately the same infinity norm, thus making it less likely to combine numbers of extreme opposite scale during the elimination process [20]. Next, let $x = D_2 y$, and solve the modified system,

$$D_1^{-1} A D_2 y = D_1^{-1} b,$$

where D_2 is a diagonal matrix that scales the columns.

$$D_1^{-1} A D_2 y = D_1^{-1} b \Rightarrow y = D_2^{-1} A^{-1} b.$$

After post-processing, we have

$$D_2 y = A^{-1} b = x.$$

Algorithm 5 outlines a basic implementation of this two-sided diagonal scaling process. LAPACK [2] uses the nearest power of 2 scaling to avoid rounding errors (see xyyEQUB routines). It is important to note that the order in which scaling is applied matters (e.g., column first scaling), resulting in different matrices with different condition numbers (see examples and discussion [18, p.45]).

Algorithm 5. (Row and Column Equilibration). Compute diagonal matrices D_1 and D_2 such that $D_1^{-1} A D_2$ has a maximum element of 1 in any row and column.

Input: $A \in \mathbb{R}^{n \times n}$
Output: $A^{(L)}$
1: $D_1 = I$
2: **for** $i = 1 : n$ **do**
3:     $D_1(i, i) = \|A(i, :)\|_\infty$
4: **end for**
5: $B = D_1 A$
6: $D_2 = I$
7: **for** $j = 1 : n$ **do**
8:     $D_2(j, j) = \|B(:, j)\|_\infty^{-1}$
9: **end for**
10: $A^{(L)} = \mathrm{fl}(\mu D_1^{-1} A D_2)$

Algorithm Performance. Algorithm performance is measured by the total number of iterations required to converge. Convergence is assessed using *Criterion 1* in [5, p.54], where the value of $\epsilon = 10^{-8}$ is set, such that

$$\|r_i\| \leq \epsilon (\|A\| \cdot \|x_i\| + \|b\|) \tag{4}$$

yielding the forward error bound

$$\|x - x_i\| \leq \epsilon \|A^{-1}\| (\|A\| \cdot \|x_i\| + \|b\|).$$

To determine the effectiveness of scaling on reducing the condition number, we approximate $\|A^{-1}\| \approx \|y\|/\|x\|$, where $Ay = x$, following [11].

Software. We utilized the *Universal Numbers Library* [44], or *Universal* for short. *Universal* is a C++ header-only template library that implements multiple machine number representations, including floats and posits, with arbitrary configurations and supports standard arithmetic operations. Clang 14 (clang-1400.0.29.202) was used to compile the code.

Hardware. All numerical experiments were conducted on a desktop computer equipped with a 3.2 GHz 6-Core Intel Core i7 processor, 32 GB of 2267 MHz DDR4 RAM, and macOS 13.5 operating system.

5   Results

The results are displayed in Table 4. The method achieved convergence across all test cases by employing Algorithm 5 to cast elements to a lower precision. However, neither Algorithms 3 and 4 guarantee convergence. In general, two-sided scaling results in the best performance using iterative refinement. We highlight the approaches that involved scaling and conversion, operating without preconditioning strategies like incomplete LU factorization or preconditioned GMRES. We found convergence with various values of μ, for example, when $\mu = 16$, Algorithm 5 converged for nos1, and with $\mu = 0.75$, Algorithm 4 converged for saylr1 and west0132, and bwm200. With an eight-more bit posit configuration, i.e., posit$\langle 24,2 \rangle$, Algorithm 3 converged for all test matrices (Figs. 3 and 4).

Table 4. Total number of iterations required by the mixed-precision iterative refinement method with scaling factor $\mu = 1/16$. Divergence is indicated by ∞. Convergence was obtained for other values of μ in Algorithm 4. For the case "nos1", indicated by '*', convergence was achieved with $\mu = 16$.

Matrix	Algorithm 3	Algorithm 4	Algorithm 5
arc130	2	1	1
bcsstk01	∞	6	4
bcsstk05	∞	5	8
lund_a	∞	27	5
nos1	∞	180	81*
pores_1	14	5	3
saylr1	∞	∞	87
steam1	∞	∞	2
west0132	∞	∞	4
bwm200	∞	63	9

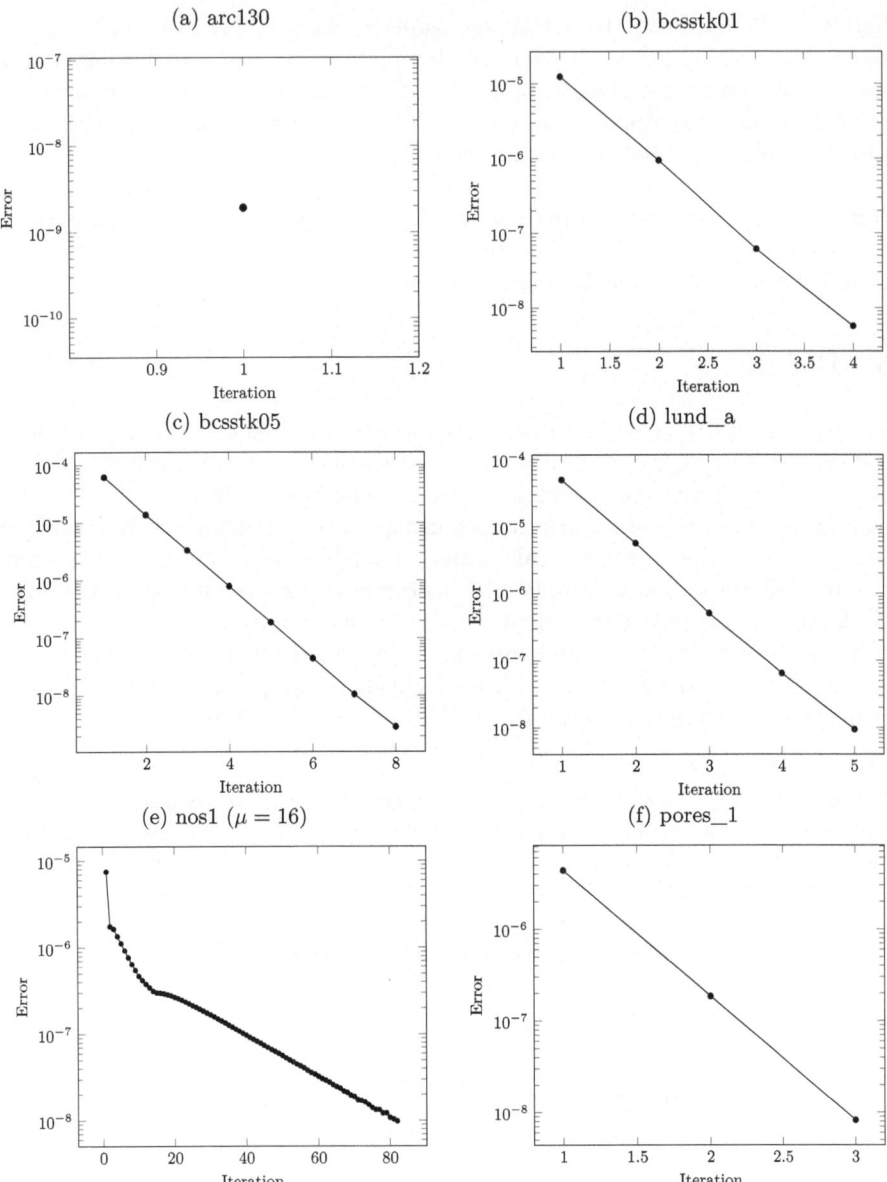

Fig. 3. Number of iterations and error using Algorithm 5 with constant $\mu = 1/16$.

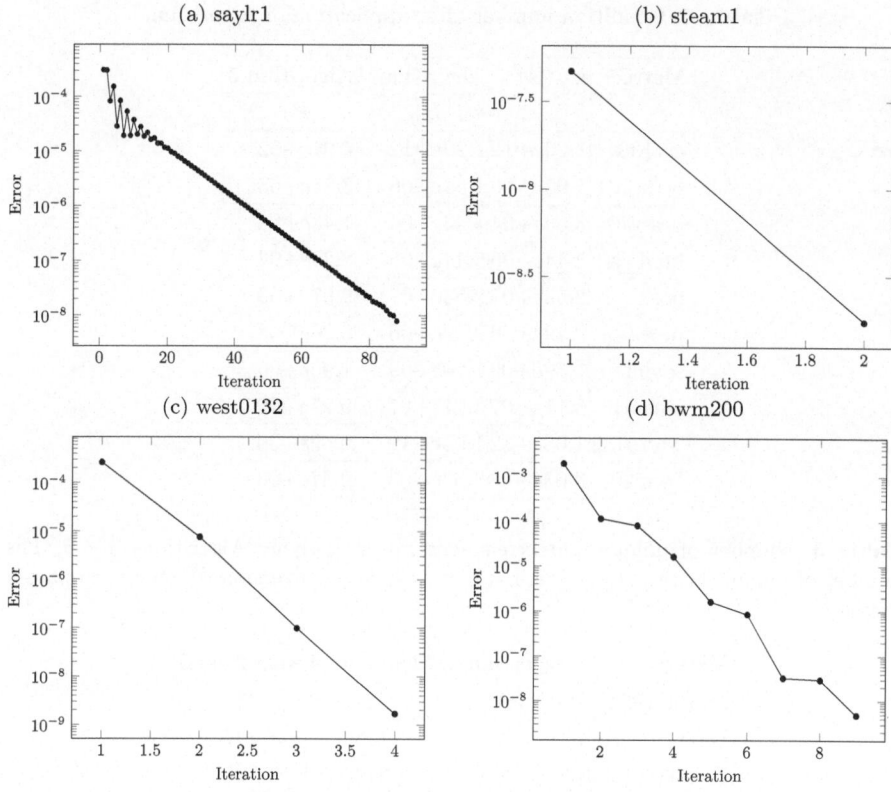

Fig. 4. Results from Algorithm 5.

Table 5 presents the reduction in condition numbers achieved by applying Algorithms 4 and 5, further investigating the effectiveness of scaling. The condition numbers shown are computed with working precision. The table shows that diagonal scaling reduces large condition numbers for the matrices under consideration. Notably, for matrices 27 and 298, the scaled matrices remain ill-conditioned. This observation highlights that convergence is not solely dependent on the condition number, as mentioned in [20]. Table 6 presents the count of distinct matrix element values both before and following the application of Algorithms 3 through 5. The table underscores how the scaling procedures employed by these algorithms influence the variability among the matrix elements.

Table 5. Condition number after application of Algorithm.

Matrix	$\kappa_\infty(A)$	Algorithm 4 $\kappa_\infty(B)$	Algorithm 5 $\kappa_\infty(B)$
arc130	1.20e+12	1.20e+12	4.07e+02
bcsstk01	1.60e+06	1.59e+06	3.25e+03
bcsstk05	3.53e+04	1.43e+04	1.43e+04
lund_a	5.44e+06	5.44e+06	8.29e+04
nos1	2.53e+07	2.53e+07	9.07e+06
pores_1	2.49e+06	2.49e+06	8.20e+03
saylr1	1.59e+09	1.59e+09	1.96e+05
steam1	3.11e+07	3.11e+07	9.27e+00
west0132	1.05e+12	1.05e+12	9.59e+06
bwm200	3.03e+03	3.03e+03	2.37e+03

Table 6. Number of unique matrix elements after applying Algorithms 3 – 5. The number of unique entries of the original matrix before conversion is denoted in parentheses.

Matrix	Algorithm 3	Algorithm 4	Algorithm 5
arc130 (961)	392	224	339
bcsstk01 (105)	15	103	232
bcsstk05 (246)	121	244	725
lund_a (232)	74	28	74
nos1 (9)	3	9	7
pores_1 (152)	102	131	132
saylr1 (705)	453	277	665
steam1 (764)	592	325	767
west0132 (237)	227	145	167
bwm200 (7)	7	7	6

6   Summary and Conclusions

We have applied mixed-precision iterative refinement using posits to find accurate approximations to $Ax = b$ for ten different sparse matrices from the *SuiteSparse Matrix Collection*. A deferred rounding mechanism in software allows us to simulate high-precision residual calculations. After applying three conversion algorithms, we find that two-side diagonal scaling leads to convergence in all cases and a significant reduction in the condition number in most cases without using a dedicated preconditioner. We found convergence comparable to similar experiments that used fp16. For most matrices in the test suite, we observed one-sided row scaling produced similar results. The convergence and convergence rate generally depend on the matrix type, condition number, and matrix size [26]. However, matrix size was not a determining factor in our experiments. Our results extend [30] to posits and [7] to nonsymmetric matrices using LU factorization instead of Cholesky.

Converting matrix entries to a lower precision for solving $Ax = b$ is a crucial yet challenging task in numerous applications, such as training deep learning models [8,39,43], image processing [12], and climate modeling [14]. In addition to the loss of precision during conversion to low precision, typically producing overflow or underflow, values get mapped to the maximum and minimum values in the range by default, which can produce undesirable behaviors. While fp16 has become a popular choice due to the increased availability of hardware support, there is a need to explore alternative formats, such as posits, in which hardware is becoming available [42]. Scaling strategies that leverage the distribution of machine numbers in the posit number system need to be further understood.

This research establishes a foundation for exploring various directions and conducting further analyses. With posits, techniques that couple iterative refinement with preconditioned GMRES and incorporate incomplete or block low-rank LU factorization, among others, need to be investigated. While optimal scaling techniques have been discussed in [6], determining when and how to scale effectively remains an open problem [16]. Various scaling strategies exist, including Hungarian scaling [31] and simple interval transformation [30], which could be explored in the context of posits. Lastly, new number systems such as *takums* developed by Laslo Hunhold offer further directions for research related to low and mixed-precision algorithms.

Acknowledgments. The first author acknowledges partial support for this research by the Maine Economic Improvement Fund (MEIF). However, the authors received no direct financial support for the research, authorship, and/or publication of this article.

Disclosure of Interests. The authors have no competing interests to declare relevant to the content of this article.

References

1. Al-Kurdi, A., Kincaid, D.R.: LU-decomposition with iterative refinement for solving sparse linear systems. J. Comput. Appl. Math. **185**(2), 391–403 (2006)
2. Anderson, E., et al.: LAPACK users' guide. SIAM (1999)
3. Baboulin, M., et al.: Accelerating scientific computations with mixed precision algorithms. Comput. Phys. Commun. **180**(12), 2526–2533 (2009)
4. Bailey, D.H., Borwein, J.M.: High-precision arithmetic in mathematical physics. Mathematics **3**(2), 337–367 (2015)
5. Barrett, R., et al.: Templates for the solution of linear systems: building blocks for iterative methods. SIAM (1994)
6. Bauer, F.L.: Optimally scaled matrices. Numer. Math. **5**(1), 73–87 (1963)
7. Buoncristiani, N., Shah, S., Donofrio, D., Shalf, J.: Evaluating the numerical stability of posit arithmetic. In: 2020 IEEE International Parallel and Distributed Processing Symposium (IPDPS), pp. 612–621. IEEE (2020)

8. Carmichael, Z., Langroudi, H.F., Khazanov, C., Lillie, J., Gustafson, J.L., Kudithipudi, D.: Deep positron: a deep neural network using the posit number system. In: 2019 Design, Automation & Test in Europe Conference & Exhibition (DATE), pp. 1421–1426. IEEE (2019)
9. Carson, E., Higham, N.J.: A new analysis of iterative refinement and its application to accurate solution of ill-conditioned sparse linear systems. SIAM J. Sci. Comput. **39**(6), A2834–A2856 (2017)
10. Carson, E., Higham, N.J.: Accelerating the solution of linear systems by iterative refinement in three precisions. SIAM J. Sci. Comput. **40**(2), A817–A847 (2018)
11. Cline, A.K., Moler, C.B., Stewart, G.W., Wilkinson, J.H.: An estimate for the condition number of a matrix. SIAM J. Numer. Anal. **16**(2), 368–375 (1979)
12. Cococcioni, M., Rossi, F., Ruffaldi, E., Saponara, S.: Fast deep neural networks for image processing using posits and arm scalable vector extension. J. Real-Time Image Proc. **17**(3), 759–771 (2020)
13. Davis, T.A., Hu, Y.: The university of Florida sparse matrix collection. ACM Trans. Math. Softw. (TOMS) **38**(1), 1–25 (2011)
14. Dawson, A., Düben, P.D., MacLeod, D.A., Palmer, T.N.: Reliable low precision simulations in land surface models. Clim. Dyn. **51**(7), 2657–2666 (2018)
15. Demmel, J., Hida, Y., Kahan, W., Li, X.S., Mukherjee, S., Riedy, E.J.: Error bounds from extra-precise iterative refinement. ACM Trans. Math. Softw. (TOMS) **32**(2), 325–351 (2006)
16. Elble, J.M., Sahinidis, N.V.: Scaling linear optimization problems prior to application of the simplex method. Comput. Optim. Appl. **52**, 345–371 (2012)
17. Feldman, M.: Fujitsu reveals details of processor that will power Post-K supercomputer (2018). Accessed 26 Mar 2019
18. Forsythe, G.E., Moler, C.B.: Computer Solution of Linear Algebraic Systems. Prentice-Hall, Englewood Cliffs (1967)
19. Golub, G.H., Van Loan, C.F.: Matrix Computations. Mathematical Sciences. Johns Hopkins University Press, Baltimore (1996)
20. Golub, G.H., Van Loan, C.F.: Matrix Computations. JHU Press, Baltimore (2013)
21. Posit Working Group: Standard for posit arithmetic. Technical report, National Supercomputing Centre (NSCC) Singapore (2022)
22. Gupta, S., Agrawal, A., Gopalakrishnan, K., Narayanan, P.: Deep learning with limited numerical precision. In: International Conference on Machine Learning, pp. 1737–1746. PMLR (2015)
23. Gustafson, J.L., Yonemoto, I.T.: Beating floating point at its own game: posit arithmetic. Supercomput. Front. Innov. **4**(2), 71–86 (2017)
24. Haidar, A., et al.: The design of fast and energy-efficient linear solvers: on the potential of half-precision arithmetic and iterative refinement techniques. In: Shi, Y., et al. (eds.) ICCS 2018. LNCS, vol. 10860, pp. 586–600. Springer, Cham (2018). https://doi.org/10.1007/978-3-319-93698-7_45
25. Haidar, A., Tomov, S., Dongarra, J., Higham, N.J.: Harnessing GPU tensor cores for fast FP16 arithmetic to speed up mixed-precision iterative refinement solvers. In: SC18: International Conference for High Performance Computing, Networking, Storage and Analysis, pp. 603–613. IEEE (2018)
26. Haidar, A., Wu, P., Tomov, S., Dongarra, J.: Investigating half precision arithmetic to accelerate dense linear system solvers. In: Proceedings of the 8th Workshop on Latest Advances in Scalable Algorithms for Large-Scale Systems, pp. 1–8 (2017)

27. He, Y., Ding, C.H.: Using accurate arithmetics to improve numerical reproducibility and stability in parallel applications. J. Supercomput. **18**(3), 259–277 (2001)
28. Higham, N.J.: Iterative refinement for linear systems and lapack. IMA J. Numer. Anal. **17**(4), 495–509 (1997)
29. Higham, N.J., Mary, T.: A new preconditioner that exploits low-rank approximations to factorization error. SIAM J. Sci. Comput. **41**(1), A59–A82 (2019)
30. Higham, N.J., Pranesh, S., Zounon, M.: Squeezing a matrix into half-precision, with an application to solving linear systems. SIAM J. Sci. Comput. **41**(4), A2536–A2551 (2019)
31. Hook, J., Pestana, J., Tisseur, F., Hogg, J.: Max-balanced hungarian scalings. SIAM J. Matrix Anal. Appl. **40**(1), 320–346 (2019)
32. Intel Corporation: BFLOAT16 - Hardware Numerics Definition (2018)
33. Kharya, P.: Tensorfloat-32 in the A100 GPU accelerates AI training HPC up to 20x. NVIDIA Corporation, Technical report (2020)
34. Kirdani-Ryan, M., Lim, K., Smith, G., Petrisko, D.: Well rounded: visualizing floating point representations (2019). https://cse512-19s.github.io/FP-Well-Rounded. Accessed 08 Oct 2023
35. Kulisch, U.: Grundlagen des numerischen rechnens-mathematische begründung der rechnerarithmetik, reihe informatik (1976)
36. Langou, J., Langou, J., Luszczek, P., Kurzak, J., Buttari, A., Dongarra, J.: Exploiting the performance of 32 bit floating point arithmetic in obtaining 64 bit accuracy (revisiting iterative refinement for linear systems). In: SC'06: Proceedings of the 2006 ACM/IEEE Conference on Supercomputing, p. 50. IEEE (2006)
37. Leon, S.J., De Pillis, L.: Linear Algebra with Applications. Pearson, London (2020)
38. Lindquist, N., Luszczek, P., Dongarra, J.: Improving the performance of the GMRES method using mixed-precision techniques. In: Nichols, J., Verastegui, B., Maccabe, A.B., Hernandez, O., Parete-Koon, S., Ahearn, T. (eds.) SMC 2020. CCIS, vol. 1315, pp. 51–66. Springer, Cham (2020). https://doi.org/10.1007/978-3-030-63393-6_4
39. Lu, J., Fang, C., Xu, M., Lin, J., Wang, Z.: Evaluations on deep neural networks training using posit number system. IEEE Trans. Comput. **70**(2), 174–187 (2020)
40. Ma, D., Saunders, M.A.: Solving multiscale linear programs using the simplex method in quadruple precision. In: Numerical Analysis and Optimization, pp. 223–235. Springer, Heidelberg (2015)
41. Ma, D., Yang, L., Fleming, R.M., Thiele, I., Palsson, B.O., Saunders, M.A.: Reliable and efficient solution of genome-scale models of metabolism and macromolecular expression. Sci. Rep. **7**(1), 1–11 (2017)
42. Mallasén, D., Murillo, R., Del Barrio, A.A., Botella, G., Piñuel, L., Prieto-Matias, M.: PERCIVAL: open-source posit RISC-V core with quire capability. IEEE Trans. Emerg. Top. Comput. **10**(3), 1241–1252 (2022)
43. Murillo, R., Del Barrio, A.A., Botella, G.: Deep pensieve: a deep learning framework based on the posit number system. Digit. Signal Process. **102**, 102762 (2020)
44. Omtzigt, E.T.L., Quinlan, J.: Universal numbers library: Multi-format variable precision arithmetic library. J. Open Source Softw. **8**(83), 5072 (2023). https://doi.org/10.21105/joss.05072
45. Palmer, T.N.: More reliable forecasts with less precise computations: a fast-track route to cloud-resolved weather and climate simulators? Philos. Trans. R. Soc. A Math. Phys. Eng. Sci. **372**(2018), 20130391 (2014)
46. Suraksha, E.P.: Bengaluru-based calligo tech to receive first silicon with posit computing capability this month. The Econmoic Times (2024)

47. Svyatkovskiy, A., Kates-Harbeck, J., Tang, W.: Training distributed deep recurrent neural networks with mixed precision on GPU clusters. In: Proceedings of the Machine Learning on HPC Environments, pp. 1–8. ACM (2017)
48. Wilkinson, J.H.: Rounding Errors in Algebraic Processes. Prentice-Hall Inc., Hoboken (1963)
49. Wu, C., Wang, M., Chu, X., Wang, K., He, L.: Low-precision floating-point arithmetic for high-performance FPGA-based CNN acceleration. ACM Trans. Reconfigurable Technol. Syst. (TRETS) **15**(1), 1–21 (2021)

Adaptive Interval Segmentation: An Algorithmic Approach for Optimal SORN Datatypes

Nils Hülsmeier[1]([✉])[iD], Moritz Bärthel[1][iD], Jochen Rust[2][iD], and Steffen Paul[1][iD]

[1] Institute of Electrodynamics and Microelectronics (ITEM.me),
University of Bremen, Bremen, Germany
{huelsmeier,baerthel,steffen.paul}@me.uni-bremen.de
[2] Hamburg University of Applied Sciences, Hamburg, Germany
jochen.rust@haw-hamburg.de

Abstract. The Sets Of Real Numbers (SORN) format is an interval-based number representation to perform fast and low-complex arithmetic operations. Since the implemented arithmetic is based on lookup tables, the applied SORN datatypes are not standardized and can be highly application specific. Because the formats precision is rather low in general, the evaluation of suitable SORN datatypes is one of the major challenges when applying the format, since not all datatypes guarantee sufficient results for the implemented algorithms. Therefore, this paper presents an algorithmic approach to determine the optimized interval distribution for a SORN datatype for specific applications. The Adaptive Interval Segmentation (AIS) algorithm is gradient based and applies directional nested intervals to adapt a floating point functionality by SORN arithmetic. This approach is used to evaluate SORN datatypes for Hybrid SORN k-Nearest Neighbor (kNN) classification. For the MNIST dataset, the AIS algorithm provides seven SORN datatypes that show better classification results for Hybrid SORN kNN classification, compared to floating point implementations. This is particular evident in a four and a five bit SORN datatype leading to an accuracy increase of 0.24% and 0.26%, respectively.

Keywords: SORN · Hybrid SORN · AIS · AIS algorithm · Adaptive Interval Segmentation · kNN

1 Introduction and Related Work

A wide range of different binary number formats is considered in recent years to improve implementations of machine learning algorithms regarding their computational costs or accuracy. Besides fixed point (FxD) or integer implementations with very short wordlengths, in order to reduce hardware complexity, also new

The authors acknowledge the financial support by the Federal Ministry of Education and Research of Germany in the project "Open6GHub" (grant number: 16KISK016).

innovative number formats such as posits and floating point number formats with untypical wordlenghs are considered [16]. Based on the IEEE-754 floating point standard [15] different resolutions and widths of mantissa and exponent are tested to find a suitable solution. This led to the brain floating point (Bfloat) format width 16 bit wordlength, but also more radical reductions are used, e.g. eight bit floating point or binary neural networks as extreme cases [16,17]. In addition to the quantization of the 32 bit floating point solution for inference, also the ability of neural networks to compensate inaccuracies of the number representation can be exploited by quantization-aware training.

The universal number (unum)-based formats represent new and innovative approaches to improve binary number representations [2,3]. While the posits are an alternative to the traditional floating point, the Sets Of Real Numbers (SORN) number format utilizes a coarse resolution to provide low complex and ultra fast computing [2]. Initially, the SORNs were designed for preprocessing in order to reduce the complexity of optimization problems by excluding wrong solutions in advance. This was used in [6,8], where SORNs are used as preprocessor for MIMO symbol detection and a sphere decoder, respectively. However, previous works have shown that by combining SORN arithmetic with FxD adder trees (Hybrid SORN approach) and the usage of appropriate datatypes, SORN-based architectures can be used in several applications. In [4] a Hybrid SORN k-Nearest Neighbor (kNN) algorithm was implemented for MNIST classification and in [7] SORNs were used to implement an edge detection. Furthermore, in [5] a Hybrid SORN hardware accelerator for Support Vector Machine (SVM) training was presented.

In contrast to FxD and floating point, SORNs are not standardized. Therefore, the reduction of the wordlength does not directly lead to a linear reduction of value range and resolution as for FxD and floating point representations. Instead, SORN datatypes provide a high degree of freedom regarding the number and the widths of the chosen intervals and the resulting resolution. Therefore, the evaluation of an appropriate datatype for a specific application is often based on experience and multiple options have to be tested in order to find a suitable solution.

Several algorithms and methods that apply interval arithmetic to determine an optimal solution for a certain problem exist. A typical numerical approach to find a root of a function is the bisection method, where intervals are bisected and the position of the root is decided based on the sign [11]. This is repeated until the optimal solution is found. This general method is also applied by the S.T.E.P. algorithm presented in [12], where the intervals are divided into two segments and the one with the lowest difficulty is chosen to find the optimum of a function for one dimension. These approaches have in common that they apply interval arithmetic to achieve an optimal solution. Hence, the interval width usually converges to zero. For SORN arithmetic, on the other hand, a set of intervals that provides a sufficient solution and has the lowest possible cardinality is required. To the best of the authors knowledge, no such algorithm exists.

In this paper an algorithm to determine an optimal SORN datatype for a specific application by using directional nested intervals is presented, referred to as Adaptive Interval Segmentation (AIS) algorithm. To validate the algorithm, it is used to determine suitable SORN datatypes for the Hybrid SORN kNN for MNIST classification, introduced in [4].

2   Hybrid SORN Arithmetic

Developed from unum type-II, the SORNs are a binary number representation that exploits a combination of half open intervals and fixed values to represent real numbers [2]. Instead of following a certain quantization or standardization as IEEE floating point and FxD, SORNs are defined by a datatype $\mathcal{D}$, which consists of specific intervals and fixed values. As an example, the five bit datatype $\mathcal{D} = \{\, 0 \ (0, 0.25] \ (0.25, 0.5] \ (0.5, 0.75] \ (0.75, 1] \,\}$ covers the value range between 0 and 1 with four intervals and one exact value. A SORN datatype is defined by its lattice values $\mathcal{L}$, which provide the information which values are used to define the intervals and exact values of the datatype, respectively. For example $\mathcal{L} = \{0, 0.25, 0.5, 0.75, 1\}$ are the corresponding lattice values for the example datatype $\mathcal{D}$. However, the lattice values do not lead to a unique SORN datatype. From the given lattice values $\mathcal{L}$ also other SORN datatypes could be built, e.g. the exact zero value could be replaced by a closed interval $\mathcal{D}_1 = \{\, [0, 0.25] \ (0.25, 0.5] \ (0.5, 0.75] \ (0.75, 1] \,\}$ or another exact value could be included $\mathcal{D}_2 = \{\, 0 \ (0, 0.25] \ (0.25, 0.5] \ (0.5, 0.75] \ (0.75, 1) \ 1 \,\}$. The lattice values and the datatypes can be freely selected, which includes lattice values $\pm\infty$ and linear, logarithmic or other datatype distributions. Hence, SORNs can be highly adapted to specific applications. In binary representation, each bit represents the presence (1) or absence (0) of the corresponding interval or exact value in the datatype. SORN arithmetic operations utilize interval arithmetic and are implemented using pre-computed lookup tables (LUTs) consisting of simple Boolean logic [6,9].

Besides fast and low complex computing, the usage of LUTs allows to implement fused SORN operations, which process multiple arithmetic operations in one LUT [10]. Another advantage is to include a functionality into a FxD to SORN conversion, such as an exponential function or a sine, instead of converting the value first and exploiting the functionality afterwards. These conversions can be necessary to implement the data transfer from the host computer to FPGA architectures with SORN arithmetic.

A disadvantage of SORN arithmetic is that additions with a non zero value are likely to result in an interval extension, caused by the usage of interval arithmetic. This can result in wide intervals, providing non-meaningful results as $[-\infty, \infty]$ in a worst case scenario. Since those results are not usable in a practical way for interpretation or further computations, for applications in high dimensional vector spaces the Hybrid SORN approach can be used. It is a combination of SORN arithmetic and FxD adder trees [4]. Once the SORN result is converted into FxD, either one or both interval bounds can be summed up by an adder tree, depending on the application the Hybrid SORN approach is used for. Hence, SORN to FxD conversions are required for the Hybrid SORN approach. These converters are typically of low complexity, resulting in an overall reduced hardware complexity for e.g. a scalar product, compared to FxD arithmetic [5].

3   Hybrid SORN k-Nearest Neighbor Algorithm

The development of SORNs in recent years allows implementations of more complex algorithms, e.g. machine learning algorithms. In particular, the Hybrid SORN approach as well as fused and functional SORN operations provide advantages in terms of hardware complexity and classification accuracy, respectively. This was shown in [4], where a Hybrid SORN implementation of the kNN was introduced.

In contrast to other machine learning algorithms like SVMs or neural networks, the kNN is not able to weight specific features stronger than others, nor to learn an optimal set of weights for all features [1]. Hence, it is a supervised machine learning algorithm that is not trained in a typical way. Instead, it determines the squared Euclidean distances between a test vector $\mathbf{x}$ and reference vectors $\mathbf{r}$ [13]. Thus, the training data is used as reference vectors for the classification. This results in the distance

$$\delta = \|\mathbf{x} - \mathbf{r}\|_2^2 \tag{1}$$

$$\delta = \sum_{i=1}^{N} (x_i - r_i)^2 \tag{2}$$

with x_i and r_i as vector entries and N as vector length. The Euclidean distance can be defined as the Minkowski distance of order two. Besides, also other distance metrics could be used for kNN classification as Minkowski distances with different orders, e.g. the Manhattan distance. However, the Euclidean distance is the most commonly used distance metric for kNN classification. For classification, the number of k smallest distances between the test vector and all reference vectors is determined. Therefore, a training dataset with M reference vectors results in M distance calculations. Based on these distances δ, the test vector is classified by the labels of the k nearest reference vectors. Since the majority of labels decides which label is classified for k larger than one, k is usually chosen as an odd value. If no majority decision can be made, the smallest distances are used for classification.

For the Hybrid SORN kNN, the squared Euclidean distance (see Eq. (2)) is determined using the Hybrid SORN approach [4]. The datapath of the Hybrid SORN calculation block of the Hybrid SORN kNN is presented in Fig. 1. Assuming that the data is stored in FxD representation, the incoming entries of test and reference vector x_i and r_i are converted into SORN, respectively (see Fig. 1, FxD2SORN block). Then they are fed into the SORN ALU, where the individual squared distances between the test vector and each reference vector $\delta_i = (x_i - r_i)^2$ are determined by a fused SORN operation. The distance is then converted into FxD (see Fig. 1, SORN2FxD block) and either one or both interval bounds are summed up by a FxD adder tree (see Sect. 2). Since the kNN determines the smallest distances, only the lower interval bound was considered for Hybrid SORN kNN in [4].

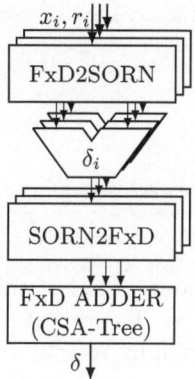

Fig. 1. Hybrid SORN calculation block of Hybrid SORN kNN to determine the Euclidean distance [4].

4   Adaptive Interval Segmentation Algorithm

The Adaptive Interval Segmentation (AIS) algorithm is presented in Algorithm 1. It searches for a SORN datatype that provides results similar to floating point calculations for a specific calculation with SORN arithmetic. For this purpose, the interval bounds of the arithmetic SORN operation are adjusted iteratively based on the gradient direction of a cost function, where the cost function is the difference between SORN and floating point result. The arithmetic operation that should be replaced by SORNs is given by a floating point functionality $f(\Upsilon)$. The target of the AIS algorithm is to find a SORN datatype that provides a solution for this functionality that matches a specific target τ, which could be the floating point result for example. Besides the functionality, its floating point inputs Υ and the target result, AIS requires lower and upper bound lb, ub of the intervals and lattice values, respectively (see Algorithm 1). The interval bounds are used as starting values for the iterative determined intervals $\mathbf{x}_{(i)}$. Based on the current interval bounds, the floating point functionality is converted to SORN functionality f_S (see Algorithm 1 step 4). Then, the cost function $c(\tau, f_S)$ with respect to the target result is determined. The direction in which the interval is changed is determined based on the gradient of the cost function for the upper and lower bound

$$\mathbf{g}_l, \mathbf{g}_u = \nabla c_l, \nabla c_u \qquad (3)$$

Algorithm 1. Adaptive Interval Segmentation

Input: target result τ, functionality $f(\Upsilon)$, inputs Υ, lower bound lb, upper bound ub
Output: lattice values $\mathcal{L}$
1: $\mathbf{x}_l \leftarrow lb$
2: $\mathbf{x}_u \leftarrow ub$
3: **for** $i = 1 \dots I$ **do**
4:      $f_S \leftarrow F2S(f(\Upsilon); \mathbf{x}_l, \mathbf{x}_u)$ ▷ Float to SORN conversion based on interval bounds
5:      $c(\tau, f_S) \leftarrow \|\tau - f_S\|_2^2$ ▷ calculate the cost function
6:      $\Delta \leftarrow \frac{\mathbf{x}_u - \mathbf{x}_l}{2}$ ▷ calculate step size for each interval
7:      calculate $\mathbf{d}(c)$ ▷ calculate direction see eq. (4)
8:      $\mathbf{x}_{(i)} \leftarrow \mathbf{x}_{(i-1)} + \text{sign}(\mathbf{d}) \cdot \Delta$ ▷ update for the next iteration
9: **end for**
10: $\mathcal{L} \leftarrow \text{unique}(\mathbf{x}_l, \mathbf{x}_u)$ ▷ save each necessary lattice value for classification once

with c_l and c_u as lower and upper bound of the SORN equivalent of the cost function (see Algorithm 1, step 4). To prevent the algorithm from running into local optima and avoid fast shrinking interval widths, only one of the interval bounds is updated per iteration and additional terms are used to determine the direction. In general, the direction d depends on the sign and the absolute value of the gradients g_l and g_u regarding each entry of the cost function. This results in six different combinations of gradients of the interval bounds to be considered:

$$c_1 : \text{sign}(g_l) = \text{sign}(g_u)$$
$$c_2 : \text{sign}(g_l) \neq \text{sign}(g_u)$$
$$c_3 : |g_l| > |g_u|$$
$$c_4 : |g_l| < |g_u|$$
$$c_5 : (g_l < 0 \wedge g_u < 0)$$
$$c_6 : (g_l > 0 \wedge g_u > 0)$$

Based on these cases, the direction can be defined as:

$$d_i(c) = \begin{cases} g_l & \text{if } (c_2 \wedge c_3) \vee (c_1 \wedge c_5) \vee (c_1 \wedge d_u \cdot \Delta > ub) \\ g_u & \text{if } (c_2 \wedge c_4) \vee (c_1 \wedge c_6) \vee (c_1 \wedge d_l \cdot \Delta < lb) \end{cases} \tag{4}$$

The direction is decisive for the determination of the updated interval. If both signs are positive, the upper bound is increased, if both signs are negative the lower bound is decreased and if g_l and g_u have different signs, the boundary with respect to the direction with a larger absolute value is adjusted.

In addition, the general conditions of the datatype, given by the interval bounds lb and ub, have to be considered. In cases the update would violate these conditions, the opposite interval bound is adjusted towards the violating bound instead. This decrease of the interval width leads to a different gradient of the opposite interval bound, which should lead to another optimal solution inside instead of outside the interval.

To include a nested interval behavior in the algorithm, instead of the actual gradient only its direction is considered, i.e. its sign. The width of the step is then only defined by the step size Δ for each interval (see Algorithm 1 step 8). It is defined as half of the current interval width with respect to each interval (see Algorithm 1, step 6). The advantage is that the solution is coarser than for an actual gradient descent and does not lead to floating point-like solutions at the cost of slower convergence.

If the functionality requires multiple dimensions of single SORN arithmetic based operations, f_S and direction $\mathbf{d}$ become matrices. For example, the kNN classification requires a sum over multiple complex functions (see Eq. (2)). Therefore, in these cases the mean of all directions regarding one classification is used to get a vector of directions. The dimension of $\mathbf{x}$ depends on the input dimension of the floating point reference functionality. It follows that for input dimension M the algorithm provides M resulting sets of lattice values. After all intervals for each functionality are determined iteratively, $\mathbf{x}$ contains all occurred intervals, which can include multiples of individual intervals. Therefore, the final set of lattice values $\mathcal{L}$ is built by including each necessary lattice value $\mathbf{x}$ for upper and lower bound once per functionality (see Algorithm 1 step 10).

From Algorithm 1 it becomes clear that the choice of the target is crucial for the AIS algorithm. If the floating point solution is given as target, e.g. the k smallest distances for a kNN (see Eq. (2)), the resulting lattice values will most likely also show a distribution and resolution similar to floating point. Since this is not practical for SORN implementations, a target should be used that is sufficient for a correct functionality. For classification tasks, it is sufficient to determine the correct class. For example the k smallest distances for kNN classification do not have to be as close as possible to the floating point distance. Instead, it is sufficient if the majority of distances is smaller than all distances with the wrong label (see Sect. 3).

The number of iterations I, required by the AIS algorithm to provide suitable solutions, strongly depends on the complexity of the functionality that should be provided by the resulting intervals. Furthermore, the size and the complexity of data for which the functionality is applied also has an effect.

5   Results

The AIS algorithm is tested for kNN classification for the MNIST dataset [14]. It consists of 70,000 28×28 grayscale images, which are separated in 10,000 test images and 60,000 training images. The full dataset contains ten different classes, while each pixel has a resolution of eight bit.

Based on the properties of the dataset, the inputs of the AIS algorithm from Algorithm 1 are chosen for MNIST kNN classification. The images are normalized, leading to $lb = 0$ and $ub = 1$ as boundary conditions and the squared Euclidean distance is used as floating point functionality $f(\Upsilon)$ for kNN. For MNIST classification with kNN, it turned out that $I = 100$ iterations of the AIS algorithm are sufficient to achieve a meaningful set of lattice values. The

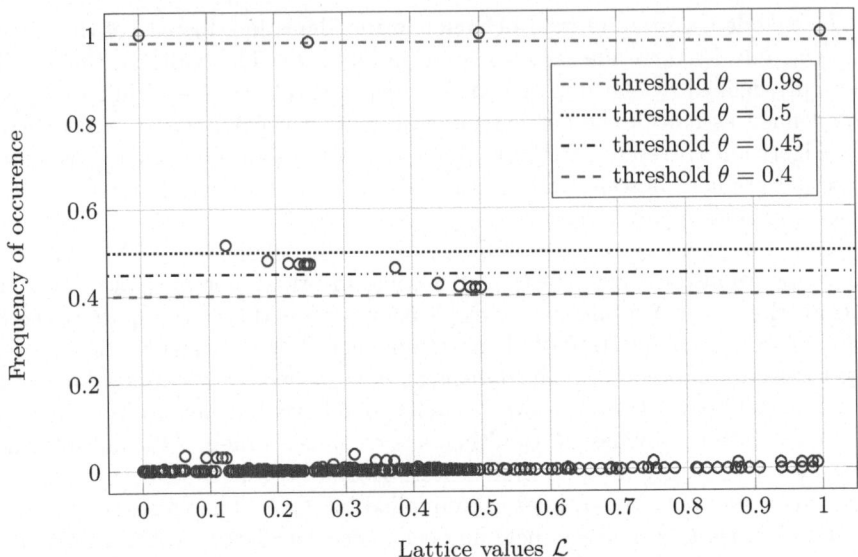

Fig. 2. Frequency of occurrence of each lattice value in all sets of lattice values determined by the AIS algorithm for kNN classification of the full MNIST training dataset and different thresholds θ for SORN datatype evaluation based on the frequency of occurrence.

distances of the images with incorrect labels minus an $\varepsilon = 10^{-3}$ are used as target value τ, as the classification is achieved by the smallest distances. Therefore, the SORN datatype has to be able to represent values that are smaller than the incorrect distances. Since the kNN is a machine learning algorithm, only the training data is used as input Υ for the AIS algorithm, while the training images with incorrect labels are used as reference vectors for kNN classification (see Eq. (2)).

This leads to 60,000 different classifications considered by the AIS algorithm to find an optimal interval distribution for kNN classification (see Sect. 4). Therefore, also 60,000 sets of lattice values are provided by the algorithm, one set for each classification task. To be usable in a practical way, one datatype for the complete dataset should be evaluated.

To evaluate one datatype for the kNN classification task, the frequency of occurrence of each lattice value in all sets of lattice values for training data classification is collected. Figure 2 shows the frequency of occurrence of all lattice values that are determined by the AIS algorithm for MNIST classification with kNN and their distribution in the value range between $lb = 0$ and $ub = 1$. It can be seen that the majority of lattice values occurs in far below 10% of the cases. Also, three values are contained in every set of lattice values, one in over 98% and another one in over 50% of the cases. Another seven lattice values show an occurrence of 45–50% and five more between 40% and 45%. The overall distribution shows higher resolution between 0 and 0.5 than between 0.5 and

1. The distribution of the most frequently occurring lattice values is showing a logarithmic shape.

Considering this distribution, several SORN datatypes could be derived. Based on the frequency of occurrence, different thresholds θ could be defined to decide which lattice values should be considered by the SORN datatype. For example four thresholds $\theta = \{0.98, 0.5, 0.45, 0.4\}$ can be derived from the distribution in Fig. 2, as described in the previous paragraph. Each of this thresholds leads to different sets of lattice values with lengths $\{4, 5, 11, 16\}$. Out of these lattice values different SORN datatypes with different distributions and lengths could be derived (see Sect. 2). When using half open intervals, it is necessary to include an exact zero value, which would increase the wordlength by one compared to considering zero by closed interval bounds for the smallest interval. Furthermore, other exact values could be included in the SORN datatype, which would increase the length of the datatype by one for each exact value. Especially for the values that occur in each determined SORN datatype this has to be discussed. However, considering the logarithmic behavior of the distribution and the arithmetic of a squared distance, this would not provide more meaningful results for most cases. As an example, the four most frequent lattice values from Fig. 2 $\{0, 0.25, 0.5, 1\}$ are compared as exact values and half open intervals in between, respectively. The square of each of these lattice values is equal to at least one of the resulting interval bounds if the half open intervals regarding the lattice values are squared. Therefore, these exact values will most likely have a low impact on the SORN LUTs and occur rarely as an exact value in the result. The exception is the exact zero, which changes the SORN arithmetic and the MNIST dataset contains multiple zeros, which leads to occurrences as one-hot encoded SORN input and output values, respectively. Therefore, it advised to consider the zero also as exact value in the SORN representations.

Besides the frequency of occurrence, also other derivations of datatypes are possible, e.g. rounding or quantization of the mean distribution or choosing the datatype based on a highly complex interval distribution for one specific classification, i.e. one particular class. However, for this application the frequency of occurrence seems to provide fairly clear results of lattice values, which are analyzed in the next section.

5.1   Classification Results

Figure 3 shows the classification results of several kNN implementations for MNIST classification. Previous works have shown that $k = 3$ provides best results for MNIST classification, e.g. in [18], which has also been confirmed for the Hybrid SORN kNN in [4]. Consequently, the target functionality in the AIS algorithm considers $k = 3$ (see Sect. 4) and the classification results in Fig. 3 are also evaluated for $k = 3$. Since the kNN determines the number of k smallest distances, the lower bound of the SORN distance calculation is considered in the FxD adder tree (see Sect. 3) for the Hybrid SORN kNN implementations.

Seven Hybrid SORN kNN implementations with different SORN datatypes evaluated by the AIS algorithm based on the frequency of occurrence with thresh-

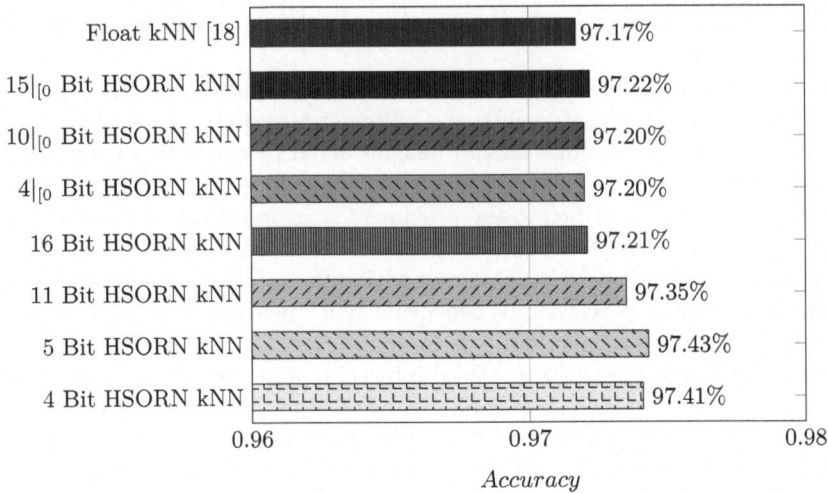

Fig. 3. Classification results of Hybrid SORN kNN and Hybrid SORN SVM implementations for MNIST dataset compared with floating point implementations.

olds $\theta = \{0.98, 0.5, 0.45, 0.4\}$, as discussed in the previous section, are compared with a floating point reference implementation from [18]. The SORN datatypes that include an exact zero are labeled with the SORN datatype length, where all intervals except the exact value zero are half open, as described in Sect. 2. The datatypes with a closed interval that includes zero instead, are labeled with $|_{[0}$. Despite the lattice value zero, the SORN datatypes are equally distributed, e.g. $4|_{[0}$ and 5 bit SORN datatypes provide the same interval distribution:

$$\mathcal{L}_5 = \{0, 0.125, 0.25, 0.5, 1\}$$
$$\mathcal{D}_{4|_{[0}} = \{\,[0, 0.125]\ (0.125, 0.25]\ (0.25, 0.5]\ (0.5, 1]\,\}$$
$$\mathcal{D}_5 = \{\,0\ (0, 0.125]\ (0.125, 0.25]\ (0.25, 0.5]\ (0.5, 1]\,\}$$

This shows that a set of lattice values of length ℓ leads to a SORN datatype of length ℓ with an exact value zero or to a SORN datatype of length $\ell - 1$ including a closed interval containing zero.

Figure 3 shows that the Hybrid SORN kNN shows better classification results for all SORN datatypes compared to the floating point implementation, which provides 97.17% accuracy for MNIST classification. The highest classification accuracies are achieved by 4 and 5 bit datatypes with 97.43% and 97.41%, respectively. The 11 bit SORN datatype provides with 97.35% a slight accuracy decrease compared to the shorter datatypes. All other implementations show classification results closer to floating point, but still with an accuracy improvement on the SORN side.

Comparisons between SORN datatypes with different zero inclusions show that the exact zero has a big impact on the classification accuracy for shorter datatypes (see Fig. 3). It improves the accuracy of the 10 and 4 bit SORN datatypes by at least 0.15%. The $3|_{[0}$ datatype does not provide meaningful results and is therefore not included in Fig. 3, but by including the exact zero it is improved to a classification accuracy 0.24% higher than floating point.

It has to be stated that the 5 bit datatype with an exact zero, which provides the best classification results, is the same datatype evaluated by traditional methods in [4]. However, the AIS algorithm provides an approach for a strategic evaluation of a suitable SORN datatype. In [4] it was also assumed that the SORN-based implementation could be an improvement of kNN classification. The results presented in Fig. 3 support this statement, furthermore it can be extended. In Sect. 3 it was discussed that the kNN algorithm is unable to increase or decrease the weight of specific features to improve the classification. However, the conversion from FxD or floating point into SORN representation and the fused arithmetic, respectively, imply a weighting of features with respect to their actual value. Considering that the datatypes are determined by applying the AIS algorithm on the training data, this approach could be understood as a training method for the kNN algorithm.

6 Conclusion

In this paper a new algorithmic approach to evaluate a suitable SORN datatype for specific applications was presented, referred to as Adaptive Interval Segmentation (AIS) algorithm. It searches for the optimal SORN datatype by directional nested intervals based on the gradient with respect to a specific target. The AIS algorithm was validated by providing suitable SORN datatypes for the Hybrid SORN kNN algorithm, resulting in seven SORN datatypes each of which provides better classification results for MNIST classification compared to 32 bit floating point kNN classification. The best Hybrid SORN kNN classification improves the accuracy by 0.26% compared to floating point, utilizing a five bit SORN representation. This improvement could be explained by the resolution of the evaluated SORN datatypes, which can be understood as a form of feature extraction. Considering this, and the fact that the datatypes where evaluated based on the training data, the AIS algorithm in combination with the Hybrid SORN kNN implementation could be understood as a training method for the kNN algorithm.

This opens a new field in research of SORN arithmetic. Therefore, in future works several topics would be of interest. First, the AIS-based datatype determination for Hybrid SORN kNN should be evaluated for other datasets. Furthermore, it would be interesting to apply the AIS algorithm to other machine learning algorithms, e.g. Support Vector Machines and neural networks, but also other algorithms that are not intentionally designed to learn could be optimized by the combination of AIS algorithm and SORN-based architectures. Finally, further studies of the AIS algorithm itself would be of interest. For example the evaluation of the datatypes out of its results or alternatives to determine the direction of the gradient.

References

1. Goodfellow, I., Bengio, Y., Courville, A.: Deep Learning. The MIT Press (2016)
2. Gustafson, J.: A radical approach to computation with real numbers. Supercomput. Front. Innovations **3**(2) (2016)
3. Gustafson, J., Yonemoto, I.: Beating floating point at its own game: posit arithmetic. Supercomput. Front. Innov. **4**(2) (2017)
4. Hülsmeier, N., Bärthel, M., Karsthof, L., Rust, J., Paul, S.: Hybrid SORN implementation of k-nearest neighbor algorithm on FPGA. In: 20th IEEE Interregional NEWCAS Conference (NEWCAS), Quebec, Canada, p. 2022 (2022)
5. Hülsmeier, N., Bärthel, M., Rust, J., Paul, S.: Hybrid SORN hardware accelerator for support vector machines. In: Gustafson, J., Leong, S.H., Michalewicz, M. (eds.) CoNGA 2023. LNCS, vol. 13851, pp. 77–87. Springer, Cham (2023). https://doi.org/10.1007/978-3-031-32180-1_5
6. Bärthel, M., Seidel, P., Rust, J., Paul, S.: SORN arithmetic for MIMO symbol detection - exploration of the type-2 Unum format. In: 2019 17th IEEE International New Circuits and Systems Conference (NEWCAS), Munich, Germany (2019)
7. Bärthel, M., Hülsmeier, N., Rust, J., Paul, S.: On the implementation of edge detection algorithms with SORN arithmetic. In: Gustafson, J., Dimitrov, V. (eds.) CoNGA 2022. LNCS, vol. 13253, pp. 1–13. Springer, Cham (2022). https://doi.org/10.1007/978-3-031-09779-9_1
8. Bärthel, M., Knobbe, S., Rust, J., Paul, S.: Hardware implementation of a latency-reduced sphere decoder with SORN preprocessing. IEEE Access **9**, 91387–91401 (2021)
9. Rust, J., Bärthel, M., Seidel, P., Paul, S.: A hardware generator for SORN arithmetic. IEEE Trans. Comput. Aided Des. Integr. Circuits Syst. **39**(12), 4842–4853 (2020)
10. Bärthel, M., Rust, J., Paul, S.: Application-specific analysis of different SORN datatypes for Unum type-2-based arithmetic. In: 2020 IEEE International Symposium on Circuits and Systems (ISCAS), Sevilla, Spain (2020)
11. Knorrenschild, M.: Numerische Mathematik, Carl Hanser Verlag GmbH & Co. KG, 5., aktualisierte Auflage (2013)
12. Swarzberg, S., Seront, G., Bersini, H.: S.T.E.P.: the easiest way to optimize a function. In: Proceedings of the First IEEE Conference on Evolutionary Computation. IEEE World Congress on Computational Intelligence, Orlando, FL, USA, vol. 1, pp. 519–524 (1994). https://doi.org/10.1109/ICEC.1994.349896.
13. LeCun, Y., Bottou, L., Bengio, Y., Haffner, P.: Gradient-based learning applied to document recognition. Proc. IEEE **86**(11), 2278–2324 (1998)
14. LeCun, Y., Cortes, C., Burges, C.: MNIST handwritten digit database. ATT Labs **2** (2010). http://yann.lecun.com/exdb/mnist. Accessed 10 Mar 2021
15. Microprocessor Standards Committee of the IEEE Computer Society: IEEE Std 754™-2008 (Revision of IEEE Std 754-1985), IEEE Standard for Floating-Point Arithmetic
16. Alsuhli, G., Sakellariou, V., Saleh, H., Al-Qutayri, M., Mohammad, B., Stouraitis, T.: Number systems for deep neural network architectures: a survey. arXiv eprint: 2307.05035 (2013)

17. Courbariaux, M., Hubara, I., Soudry, D., El-Yaniv, R., Bengio, Y.: Binarized neural networks: training deep neural networks with weights and activations constrained to +1 or −1. arXiv eprint: 1602.02830 (2016)
18. Grover, D., Toghi, B.: MNIST dataset classification utilizing k-NN classifier with modified sliding-window metric. In: Arai, K., Kapoor, S. (eds.) CVC 2019. AISC, vol. 944, pp. 583–591. Springer, Cham (2020). https://doi.org/10.1007/978-3-030-17798-0_47

Unleashing Simple Pendulum Dynamics with Posit Arithmetic

Avinash Aldhapati, Ashwini Jaya Kumar, and Rajaraman Subramanian[✉]

Calligo Technologies Pvt. Ltd., Bengaluru, India
{avinash.aldhapati,ashwini.jayakumar,
rajaraman.subramanian}@calligotech.com

Abstract. Simulations provide a powerful means to explore, analyze, and understand complex systems, allowing us to make informed decisions across various domains. Fields like genomics, physics, and climate science and many more heavily rely on simulations to produce valuable information but the Memory Wall, the gap between computation speed and data access, poses challenges, especially these days when massive amount of data is produced and quick access to it is of high priority. While solutions exist, they often require infrastructure changes or apply to specific algorithms only. Posit, a novel datatype, matches standard sizes but with higher precision. Analyzing systems like a pendulum with posits reveals lower errors compared to floats, enhancing accuracy within memory constraints. Posits excel in capturing system dynamics, surpassing competitors of similar sizes. They enable better predictions of complex simulations, advancing our understanding of natural phenomena while addressing memory limitations.

Keywords: Posits · IEEE 754 · Simple Pendulum · Simulations · high-performance computing · Error · Precision

1 Introduction

In the realm of scientific exploration and computational research, simulations have emerged as invaluable tools for understanding complex phenomena and solving intricate problems. By mimicking real-world processes in a controlled virtual environment, simulations offer researchers a means to gain insights, make predictions, and optimize systems in various domains [4,5,21]. They allow scientists and engineers to model intricate systems, such as climate patterns [6], fluid dynamics [7], molecular interactions [8], and economic markets [9], to name a few.

Simulations provide a cost-effective and time-efficient way to study phenomena that are otherwise difficult, dangerous, or expensive to observe directly. By creating virtual environments, researchers can experiment, explore different scenarios, and analyse the outcomes to gain valuable insights and drive innovation. However, simulations often deal with massive amounts of data, posing significant challenges in terms of storage, processing, and analysis. As scientific problems become increasingly complex, traditional computing architectures struggle to cope with the demands imposed by the ever-growing data volumes.

To address the storage and computational demands of simulations, various methods have been developed, including parallel processing and distributed networks [10–12]. Parallel processing techniques involve breaking down complex computations into smaller tasks that can be executed simultaneously on multiple processors, thus accelerating the overall computation. Distributed networks, on the other hand, leverage interconnected systems to distribute the computational load across multiple machines, enabling efficient processing of massive datasets. While parallel computation and distributed networks are utilized, it is important to note that algorithms are rarely entirely parallel. Consequently, there remains a necessity to transfer small portions of memory back and forth multiple times. Despite the use of parallel processing and distributed architectures, the nature of many algorithms necessitates intermittent data transfers, which may introduce additional latency and overhead. So, even with advanced computing methods, the memory wall [13, 14, 22] problem remains a significant bottleneck. As Moore's Law, which states the doubling of transistor density every two years, reaches its limitations, the focus shifts from increasing processor speed to enhancing other aspects of computing performance. One critical aspect is the memory bandwidth, the rate at which data can be read from or written to memory. The memory bandwidth often struggles to keep up with the speed of processors, creating a significant performance gap.

One potential solution to alleviate the memory wall problem is the adoption of custom data types with smaller sizes tailored to specific simulations. Custom data types can significantly reduce the volume of data required for computations, optimizing memory usage and increasing computational efficiency. However, the adoption of custom data types comes with its own challenges, particularly in legacy software systems that rely on standard data types. The introduction of non-standard data types could break compatibility and require extensive software re-engineering.

Fundamentally, we are seeking a data type that satisfactorily meets two key conditions.

- It should adhere to the IEEE size guidelines to ensure compatibility and interoperability across various systems.
- It should provide significantly greater precision than standard IEEE data types.

These requirements align precisely with the intended purpose and claims of the *posit* data type.

In order to assess the effectiveness of using Posits in scientific computations, especially in areas involving physics simulations we employ them in simulating a simple pendulum and keep track of how various dynamic variables behave. We analyse various states resulting from a wide combination of parameters pertaining to the simple pendulum and notice that Posits perform much better than the standard IEEE floats across all states.

The main contributions of the paper are:

– Simulation of simple pendulum with 64-bit IEEE floats (*double*), 32-bit IEEE floats (*float*) and 32-bit *posits* to show variation in parameter characteristics.
– Demonstration of the advantages of using posits instead of IEEE floats to simulate simple physical phenomena.
– Simulation using *posits* is done on real hardware supporting *posit* operations.

2   Background

2.1   Number System

IEEE 754 Single Precision. The IEEE *float* and *double* representations are widely used formats for representing real numbers in computer systems, adhering to the standards defined by IEEE 754 which specifies the formats for floating-point numbers, defining both single-precision (*float*) and double-precision (*double*) representations.

Few shortcomings of IEEE 754 [17, 25] are:

– IEEE 754 makes use of overflow (accepting ∞) as a substitute for large numbers and underflow (accepting 0) as a substitute for small magnitude zero numbers.
– There is additional computational overhead while dealing with denormalized numbers which have reduced precision.
– Identical computations can lead to multiple results across different computing platforms.
– Rounding is performed on individual operands of every calculation.
– The IEEE representation allows large exponents to have higher precision which is not practical.
– Wasted bit patterns; Negative Zero, multiple NaN values

To overcome all the above disadvantages of Floating point representation, in 2017, Posit [1], a novel number format was introduced, claiming to provide more accurate answers with an equal or relatively smaller number of bits and simpler hardware compared to floats.

Posit Number System. The Posit number representation is a novel framework that offers an alternative to the traditional floating-point format for numerical computation. Developed by John Gustafson [1], it aims to address some of the limitations of floating-point arithmetic, such as precision loss and rounding errors [3].

Quire. The Posit arithmetic conventions emphasize the utilization of a large register known as a *"quire"* [26] for performing collective operations, such as the dot-product. This approach allows for the deferral of rounding until the completion of the entire calculation. This is extremely beneficial when dealing with numerical integration where a large number of iterative operations have to be performed for accurate results but at the same time lead to unwanted errors due to multiple rounding offs.

For *posit*<32, 2> and *posit*<64, 3> the *quire* size is 512 *bits* and 1024 *bits* respectively [26].

3   Theoretical Details and Methodology

3.1   The Simple Pendulum

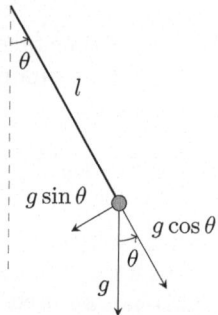

Fig. 1. A simple Pendulum

The simple pendulum is a classic example of a physical system consisting of a mass (known as the bob) attached to a string or rod of negligible mass (Fig. 1). When the bob is displaced from its equilibrium position and released, it swings back and forth under the influence of gravity, exhibiting periodic motion.

The simple pendulum provides an excellent example for understanding fundamental principles of physics and serves as a simple yet powerful mathematical model for understanding and analyzing various oscillatory systems. It provides a basis for exploring concepts like period, frequency, amplitude, and resonance, which are of key interest in physics.

Although the simple pendulum may appear elementary, it has practical applications in various domains such as in mechanical systems, electrical circuits, chemistry and spectroscopy. The behavior of a simple pendulum can be described using the principles of classical mechanics. The motion of the pendulum is primarily governed by the relationship between the length of the string, the mass of the bob, and the acceleration due to gravity. Additional parameters such as damping can also be incorporated to allow for resistance that might arise due to air and other factors. Assuming point mass, the simple pendulum system can be represented by Eq. 1 [23].

$$\frac{d^2\theta}{dt^2} + \frac{g}{l}\sin\theta + q\frac{d\theta}{dt} = 0 \;, \tag{1}$$

$\theta \rightarrow$ angle between the vertical and

the rod/string connecting the bob

$\frac{d^2\theta}{dt^2} = \alpha \rightarrow$ is the angular acceleration

$q \rightarrow$ damping coefficient

$\frac{d\theta}{dt} = \omega \rightarrow$ angular velocity

$g \rightarrow$ acceleration due to gravity

$l \rightarrow$ length of string/rod

The period of the pendulum, which is the time taken for one complete oscillation, is primarily determined by the Eq. 2 [23], assuming no damping.

$$T = 4\sqrt{\frac{l}{g}} \int_0^{\frac{\pi}{2}} \frac{d\varphi}{\sqrt{1 - k^2\sin(\varphi)}} \tag{2}$$

The Integral in Eq. 2 is an elliptical integral of the first kind, which does not have a closed form. So the possible ways to find the solution involves a series expansion or a numerical one. This can be used as a metric to determine and compare the accuracy of a various number formats. We have used IEEE 32 bit *float* and *posit*<32, 2> to determine the time period of the simple pendulum for various angles and analysed them. The results show that the Posit number format performs much better than the IEEE float.

The motion of a simple pendulum is approximately simple harmonic, meaning that the displacement follows a sinusoidal pattern. Both the theta(θ) and omega(ω) graphs are almost sinusoidal. An ideal behavior of a simple pendulum is shown in Fig. 2.

Simulating a simple pendulum may seem easy, but when there is high resolution computation involved it becomes challenging, especially due to round off errors which result in anomalous behaviour [20]. Before diving deeper into the underlying factors contributing to this occurrence it is necessary to understand the simulation algorithm that is used.

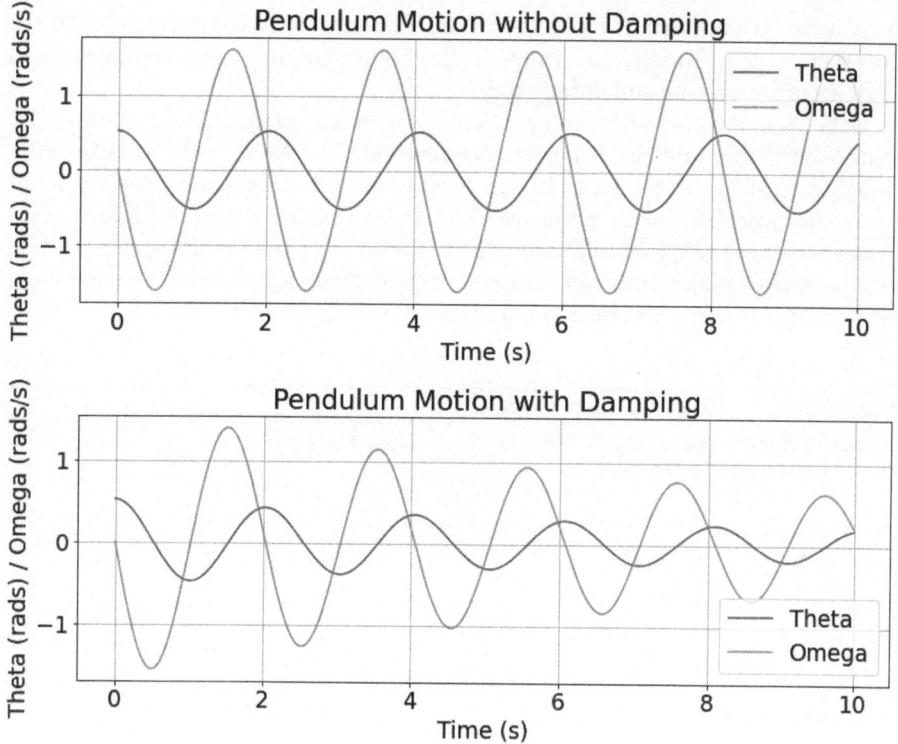

Fig. 2. Behaviour of an Ideal Pendulum

3.2 Euler-Cromer Method

The Euler-Cromer method is a numerical integration technique used to solve ordinary differential equations [24]. It is similar to the Euler method but incorporates a small modification by updating the velocities using the new positions before calculating the new positions themselves. This method is preferred over the regular Euler method because it provides more accurate results, especially for oscillatory systems(as it shows better stability for conservative systems), while maintaining computational simplicity.

Using the Euler-Cromer Method to simulate a simple pendulum would involve the following equations:

$$\alpha = - \left(\frac{g}{l} \right) \sin \theta_i - q\omega_i \ , \tag{3}$$

$$w_{i+1} = w_i - \alpha \Delta t \ , \tag{4}$$

$$\theta_{i+1} = \theta_i + w_{i+1} \Delta t \ , \tag{5}$$

Δt plays a very important role in determining how close the simulation is to the true system [24]. Smaller the value of Δt, higher the temporal resolution and more accurate are the simulated values.

Other popular numerical integration methods include the Euler method like mentioned before and the Runge-Kutta method. The Euler method is the simplest but least accurate, while Runge-Kutta methods offer higher accuracy but are computationally more intensive. The Euler-Cromer method Algorithm (1) strikes a balance by providing improved accuracy over the regular Euler method without introducing significant computational complexity, making it a commonly used choice for many physical simulations and computational models.

Algorithm 1. Euler Cromer Method
___
1: **procedure** EULER-CROMER(θ_{init},ω_{init},l,q,Δt,g,t_{init},t_{fin})
2:    $t_i \leftarrow t_{init}$
3:    $\theta_i \leftarrow \theta_{init}$
4:    $\omega_i \leftarrow \omega_{init}$
5:    **while** $t_i <= t_{fin}$ **do**
6:        $\alpha \leftarrow \left(\dfrac{g}{l}\right) \sin\theta_i - q\omega_i$
7:        $\theta_{i+1} = \theta_i + w_{i+1}\Delta t$
8:        $w_{i+1} = w_i - \alpha\Delta t$
9:        $t_{i+1} \leftarrow t_i + \Delta t$
10:    **end while**
11:    $\theta_{fin} \leftarrow \theta_i$
12:    $\omega_{fin} \leftarrow \omega_i$
13:    **return** $\theta_{fin}, \omega_{fin}$
14: **end procedure**

---

4   Related Work

In [1], John L. Gustafson leverages Goldberg's Thin Triangle problem as a means to showcase the superior accuracy offered by Posits when compared to the IEEE standard floats. By comparing the results obtained from a 128-bit IEEE *float* and a 128-bit *posit*, it is demonstrated that the *posit* format exhibits an 1.8x decimal digits accuracy increase compared to *floats* for the specific test.

The numerical stability of *posits* is evaluated by solving linear systems using the Conjugate Gradient Method as an iterative solver in [15]. Specifically, the paper investigates the potential improvement in accuracy that can be achieved by employing a *posit*<32, 2> against the IEEE 32-bit *float*. The results of the study demonstrates that when matrices fall within their native range, there is generally no significant difference in numerical accuracy but when the same matrix is re-scaled to be optimized for the *posit* representation, the *posit*<32, 2> format can provide up to 4 additional bits of precision compared to float.

The inclusion of the *quire* is discussed in [16]. Which is a huge accumulator and an integral part of the posit standard, introducing new possibilities and could potentially render certain floating-point techniques obsolete if the quire can provide equivalent functionality at a reduced cost. When performing summations involving smaller terms that result in an output with an exponent close to zero, *posits* are expected to exhibit greater accuracy compared to floats of the same size and do even better if allowed to leverage *quire* which round off only once when converted to posit.

In [19], a math library is specifically designed for posits, utilizing the CORDIC method where the implementation of trigonometric functions in a 32-bit *posit* format yields greater accuracy compared to a 32-bit *float* implementation. These results highlight the enhanced precision achievable with posits when employing the CORDIC method for trigonometric calculations.

Finally, [18] explores how the use of 16-bit *posit*, specifically with one exponent bit, significantly improves the performance and accuracy of the Lorenz 1963 model when compared to 16-bit half-precision *float*. The results also indicate that 16-bit *posit* outperforms 16-bit *float* and exhibit great promise for high-performance computing in Earth System modeling dealing with 2D fluid flows [28,29].

Although few papers such as [15] mention *quire*, they perform their analysis without utilizing them. Now that *quire* has become an integral part of the *posit* standard, it is crucial to observe the effect they have on the results relative to not using them at all. In studies that do utilize them, they have shown to give better results than using *posit* only [2,19].

It is also important note that most papers dealing with applications of posits use a software emulation to implement them while we perform the simulation on true posit based hardware.

5   Implementation

5.1   Posit Hardware

Posit used in the simple pendulum simulation is not emulated. We employed true *posit*-based computations utilizing tangible hardware capable of executing *posit* operations. The simulation was run on Calligo's UTTUNGA(v1.0) [27] - a PCIe add on card powered by TUNGA(v1.0), an SoC that houses the PNU (Posit Numerical Unit), an octa-core processor with *posit* support in each core. Although an add on card, UTTUNGA can perform computations independently, as it contains it's own operating system. The GCC compiler(v.12) is modified to compile C/C++ codes to run on posit hardware. Standard IEEE data types mentioned in the C/C++ codes such as *float* and *double* are implicitly converted to *posit*<32, 2> and *posit*<64, 3> respectively. All math functions belonging to the *math.h* header are also supported in the hardware. The outputs of the simulation is redirected to a CSV file for further analysis. This is possible on this novel hardware, as the GNU toolchain has been modified to use *posits*, making standard operating system services available.

5.2 Approach

Before delving directly into quantitative analysis, it is crucial to understand how *posits* and *floats* can impact simulations. We do this by tracking the values of important parameters like time(t), theta(θ) and omega(ω). We need to assess whether round-off errors, resulting from *float*'s and *posit*'s limitations have a significant impact on the state of the pendulum.

Although no simulation can provide exact values for real phenomena, we attempt to achieve a close approximation, especially with constrains such as limited memory. By conducting a thorough analysis of the effects of posits on the simulation and carefully considering the aforementioned factors, we can gain valuable insights into the feasibility and accuracy of using posits as an alternative to floats. To ensure a comprehensive study of the simple pendulum's behavior, it is necessary to consider different values of the initial angle. However, examining the pendulum's behavior in a detailed manner for all theta(θ) values is not practical. Therefore, for the sake of convenience and a balanced perspective, we select a reference theta(θ) value of 90°, which lies midway between the extreme values of 0 and 180°.

A small value of Δt is preferred, as it leads to more accurate simulations. For *float*, the maximum temporal resolution is achieved by utilizing all of its significant digits (approximately 7). Assuming one digit for the integer part leaves us with 6 decimal digits. Therefore, we set Δt to 10^{-6} ensuring the highest temporal resolution possible. The same value is used for *posit* as well.

Determining the appropriate observation time for the pendulum's behavior is crucial. We pick a single oscillation as the observation period for the sake of uniformity. Therefore, the simulation should cease when the theta(θ) value starts off at its initial value, crosses the x-axis and returns back to the initial angle. However, in floating arithmetic, achieving precisely the same value might not always be possible. For instance, the pendulum may return to a value of 89.999876° instead of exactly 90°. Therefore, it is essential to define a range of values within which, if a theta(θ) value falls, the simulation terminates. We must make sure that the range is always greater than or at least equal to Δt. If Δt exceeds the range, there is a possibility that the entire range gets skipped, and the simulation time will exceed that of a single period. So, any errors in parameters such as theta(θ), omega(ω) and alpha(α) or incorrect values of Δt and range directly affect the time period. For example, if Δt is 0.001 and the range is 0.0001, with initial time as 2.0001 s and initial theta(θ) as 1.5068 rad, the initial theta value (θ_i) is 1.5070 rad. If the theta(θ) value falls within the range of 1.5070 ± 0.0001, the simulation should stop. However, if Δt is larger than the range, the $\Delta\theta$ value calculated as $\Delta t * \omega$ may be significant enough that $\theta_i + \Delta\theta$ gives rise to a value greater than 1.5071, skipping the entire range. We will discuss the consequences of selecting different Δt and range values later. However, note that Δt should never be larger than the range, and the range should be as small as possible to achieve more accuracy. For our simulation, we will consider both Δt and the range to be 10^{-6} s.

The simple pendulum simulation is primarily conducted using 3 data types - *double*, *float* and *posit*. *Posit* with *quire* is also used to observe the advantages it brings forth.

6   Results and Discussion

The graph in Fig. 3 depicts the relationship between theta(θ) and time(t) for a single cycle of a simple pendulum. The pendulum has an initial angle of $90°$ and a length of 1 m. The graph is zoomed in at increasing scales, focusing on a time range close to 1.18 s. The purpose is to observe how different data types and number representations affect the behavior of the simulated pendulum. Upon closer examination, it becomes evident that certain lines in the legend, such as the '*double*,' are initially obscured by other similar curves. However, as we zoom in further, we begin to distinguish these lines. Figure 3a already demonstrates a noticeable difference between the curves of float and *posit*<32, 2>. The extent of deviation will be investigated later. For now, let's shift our attention to Fig. 3.b which provides a magnified view of Fig. 3.a.

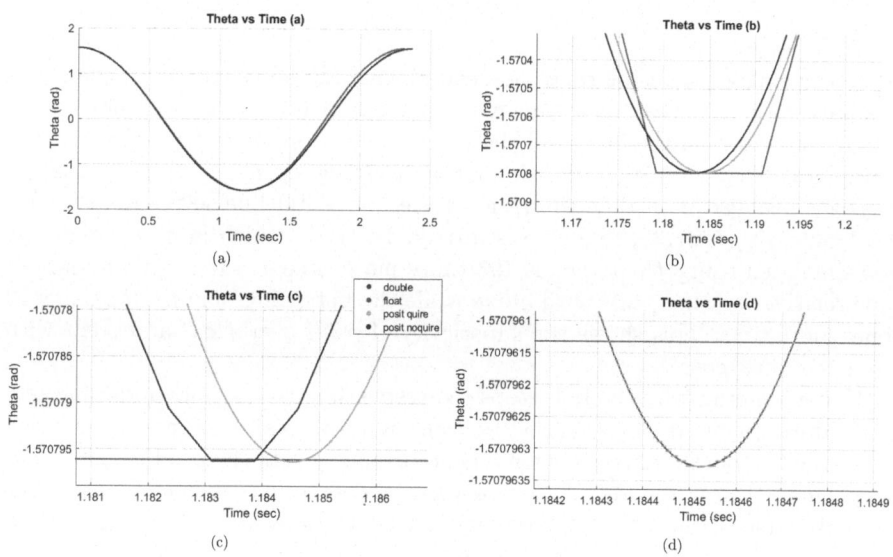

Fig. 3. Theta vs Time Plot

In Fig. 3.b the float curve struggles to approximate the true curve and instead resembles a polyline(a curve with multiple straight lines). It is crucial to note that the 'float' polyline contains a horizontal line segment, indicating a constant theta(θ) value for a certain time range. Physically, this implies that the pendulum's bob comes to a halt during that time interval. In an ideal simple pendulum, the bob momentarily stops when it reaches its maximum amplitude ($-90°$ or

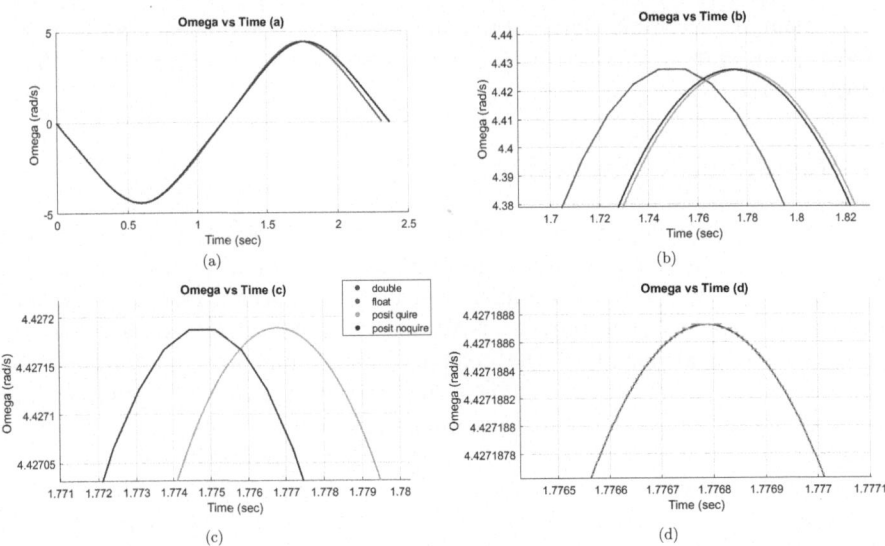

Fig. 4. Omega vs Time Plot

90°). The smaller the time range associated with the bob's stoppage, the closer it aligns with the ideal condition. One can comment on the accuracy of the simulation based on the length of the horizontal line segment from the Fig. 3.b, in which case we note that the simulation accuracy of *double* > *posit*<32, 2> > *float*. We also note that the ratio of the horizontal line segments of *float*: *posit*<32, 2>: *posit*<32, 2>_*quire* is around 193:12:1, which indicates that the bob simulated using *float* spends 193 times more time at the maximum amplitude relative to using *posit* with *quire* while the bob using only *posit* spends 12 times more time. This shows the superiority of *posit* over *float* and *posit* with *quire* over just *posit*.

Understanding why the bob stops (corresponding to the horizontal line segment) during simulations is crucial, as it also impacts *posits*, as shown in Fig. 5.c There are two potential reasons for this occurrence. First, the product $\Delta\theta$ might be so small that it rounds off to zero when added to θ_i, resulting in the same theta(θ) value. Alternatively, there may not be an equivalent representation for the new theta value (θ_{i+1}), causing it to round off to the previous theta(θ) value. This process continues until $\Delta\theta$ becomes large enough that the new theta value (θ_{i+1}) does not round off.

Let us also take a look at the region when the curve reaches its maximum or minimum value. Upon examination of the graph from Fig. 3, it is apparent that the '*float*' polyline reaches its minimum before 1.18 s, whereas the *double* does so at around 1.1845 s. Consequently, the pendulum simulated using *float* completes its oscillations much earlier than it should ideally. While *posit* without *quire* also shows this behaviour, it happens at around 1.183 s much closer to the true value at which the pendulum reaches its minimum.

Posit with quire however captures the true behaviour of the pendulum albeit with discrete steps coinciding with the true curve. This advantage of *posit* can be associated to its ability to represent smaller values much accurately than *float* and round offs at every single operation can be avoided by using *quire* which allows for controlled round offs when required.

Omega also behaves similarly. The graph of omega from Fig. 4 also reveals the previously discussed horizontal line, indicating constant angular velocity during a time interval. This is another reason why incorrect theta(θ) values are produced. Omega(ω) value rounding off occurs due to angular acceleration, similar to how theta(θ) values are affected by omega(ω) values. As angular acceleration is a function of theta(θ), a positive feedback loop emerges, leading to erratic behavior in the simulation over time. This indicates how rounding off a single variable can disrupt the ideal behavior of a system when there is significant inter-dependency between variables.

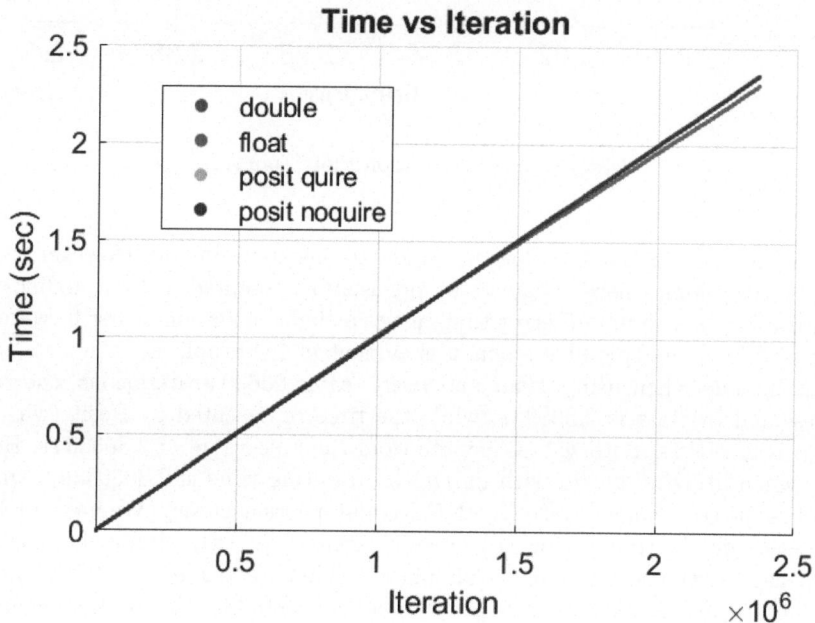

Fig. 5. Time vs Iteration Plot

One variable that remains entirely independent is time(t). Any round-off errors, if present, are solely attributable to the data types employed. Analyzing these errors is important, as regardless of the system being simulated, whether it be a pendulum or fluid flow, time-dependency plays a vital role, making it susceptible to the effects of rounding off.

Figure 5 and Fig. 6 illustrate a time vs. iteration graph, where at each iteration, Δt is added to the previous time (t_{i-1}) to calculate the current time

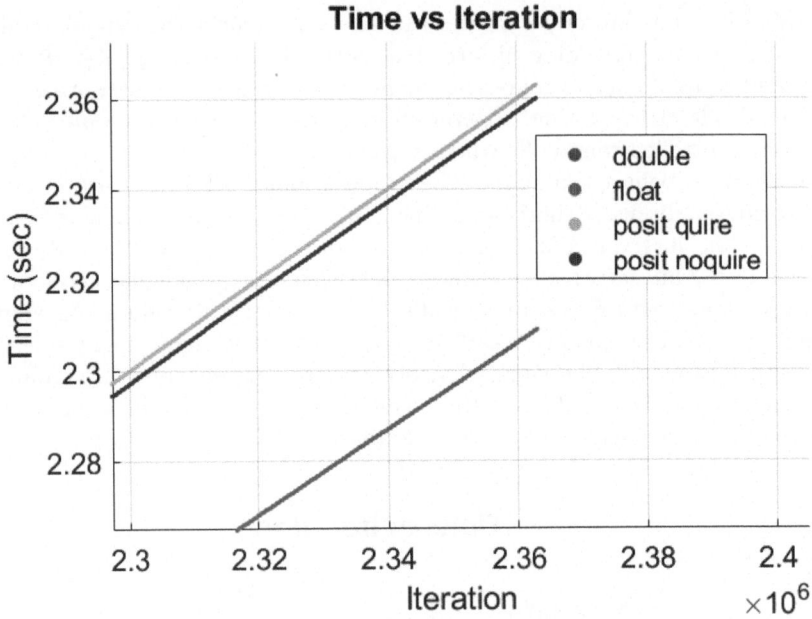

Fig. 6. Time vs Iteration Plot (Zoomed)

(t_i). As Δt remains fixed, it does not undergo any rounding off. However, there may be situations where t_i cannot be precisely represented in the number system, leading to a round-off error that causes a slight deviation either forward or backward in time. This phenomenon is evident in the graph, as it depicts variations in time when using *float* and *posit*. After 2363040 iterations, the time represented by float is 2.30893 s, while the time represented by Posit (without quire) is 2.36008 s. Both values deviate from the true value of 2.363034 s. However, when utilizing *posits* with *quire*, the resulting time is 2.3630339 s, which matches the true value (*double*) with 5 decimal places accuracy. We have already discussed how the time period (T) serves as a valuable metric, if not the sole one, for measuring the accuracy of the simulation. This is attributed to the fact that the time period is influenced by all parameters, including theta(θ), omega(ω), Δt, l, g and range.

Table 1 lists time periods for various values of theta while maintaining Δt and range at 10^{-6} s. We note that across all theta values posits perform much better than *float* and *posit* with *quire* perform exceedingly better than posits matching on an average, up to 5 decimal places with the true values. This is significant as the time period takes into consideration all errors accumulated over a time period. *Posits* with *quire* show their robust nature by withstanding all such numerical errors. Double values are rounded off and we see either 999 or 000 at the end as compared to Posits.

Table 1. Time Period of a Simple Pendulum for various values of Theta

Theta	double	float	posit-no-quire	posit-quire
10	2.009832999	1.968970299	2.007442906	2.009832993
20	2.021708999	1.979920387	2.019272685	2.021708995
30	2.041390999	1.997033119	2.038944721	2.041391000
40	2.069428999	2.024882317	2.066957205	2.069427996
50	2.106491999	2.060842514	2.103945106	2.106491997
60	2.153485999	2.102946281	2.150790989	2.153485998
70	2.211636999	2.15907383	2.208842664	2.211636990
80	2.282593999	2.226900101	2.279701530	2.282591998
90	2.368596999	2.308929443	2.365575000	2.368595987

Table 2. Time Period of a Simple Pendulum for different values of Δt and range (90°)

Δt	range	double	float	posit
0.000001	0.000001	2.36859699	2.308933258	2.368595987
0.000001	0.00001	2.36761999	2.308860779	2.367619991
0.000001	0.0001	2.36453099	2.308140755	2.364530995
0.000001	0.001	2.35476299	2.300941467	2.354763001
0.000001	0.01	2.32387299	2.271541595	2.323872998
0.000001	0.1	2.22616799	2.178404808	2.226167991
0.00001	0.00001	2.36761000	2.370437384	2.367609933
0.00001	0.0001	2.36452000	2.367353201	2.364519938
0.00001	0.001	2.35475000	2.357579947	2.354749947
0.00001	0.01	2.32386000	2.326647997	2.323859944
0.00001	0.1	2.22616000	2.228815317	2.226159945
0.0001	0.0001	2.36440000	2.364345074	2.364400029
0.0001	0.001	2.35470000	2.35455513	2.354700028
0.0001	0.01	2.32380000	2.323686838	2.323800027
0.0001	0.1	2.22610000	2.226087093	2.226100027
0.001	0.001	2.35399999	2.354011774	2.353999972
0.001	0.01	2.32299999	2.323014021	2.322999969
0.001	0.1	2.22499999	2.225021124	2.224999979
0.01	0.01	2.30999999	2.309998274	2.310000002
0.01	0.1	2.21999999	2.21999836	2.219999998
0.1	0.1	2.10000000	1.34798646	2.100000008

Table 2 is also provided to remove any suspicion that the *posit* values are sensitive to Δt and range.

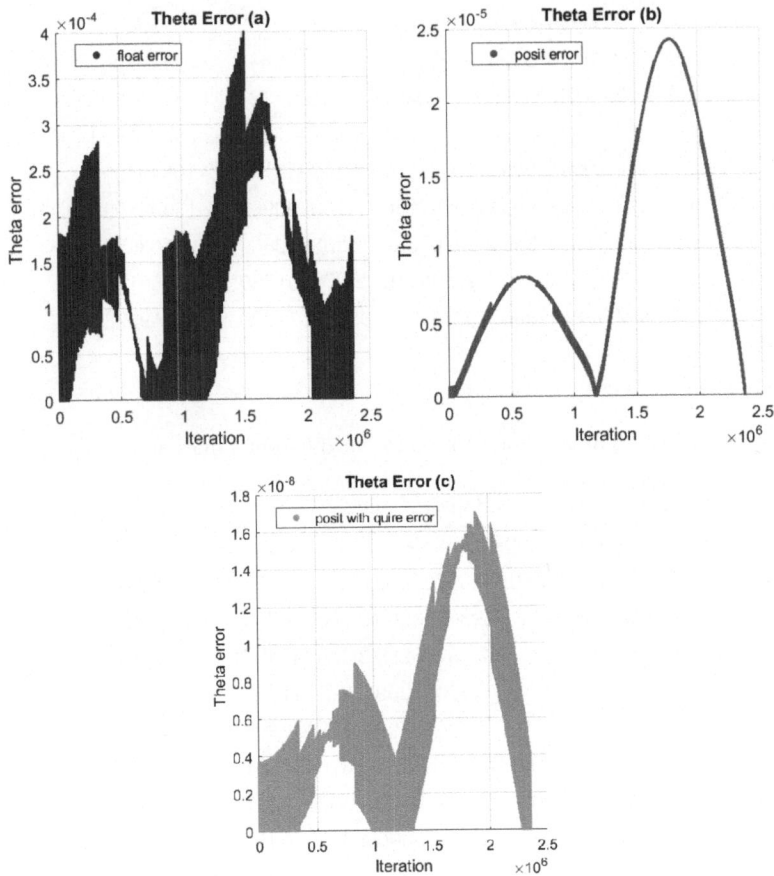

Fig. 7. Theta Error vs Iteration

In Fig. 7, the error graph of theta(θ) is depicted. Note that all error metrics are with respect to double, i.e., the "true" value and are absolute errors. Which is why we see only three curves instead of four. Although the final value after a single time period (T) offers an understanding of the overall deviation from the true value, it is equally essential to comprehend the dynamics of the error within a time period. Which is why we take a look at the error of theta(θ). The most notable feature of the graph is the sharp edges of the error curves resembling a fuzzy, thick, solid line which is actually a densely packed high frequency function.

To explain this nature of the curve, we can turn to Weierstrass approximation theorem, which states that a finite curve can be approximated by a series of straight lines. Figure 8 provides a visual representation of this concept. As

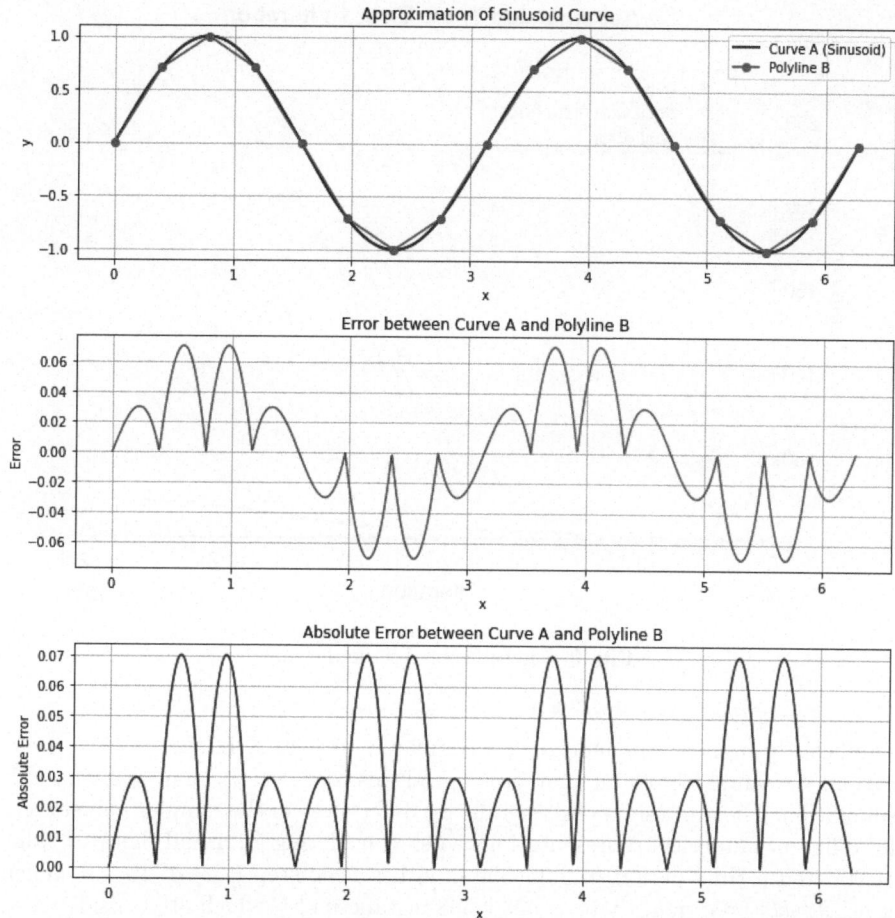

Fig. 8. Demonstration of Weierstrass theorem.

the number of line segments in the poly-line used for the curve approximation increases, the approximation becomes closer to the true curve, resulting in
reduced error amplitudes. This is evident from Fig. 3.c and Fig. 3.d where *posits*
demonstrate a better approximation of the true curve, leading to lower error
amplitudes compared to *float*. From Fig. 9, the gradual accumulation of error
also becomes apparent as the second half of the time period exhibits a higher
absolute amplitude, indicating a continuous accumulation of error that might
significantly deviate from the true curve in the long run. While it is apparent
that error occurs for both *float* and *posit*<32, 2>, *float* exhibits relatively larger
errors. However, it is important to note that this observation does not hold until
we measure the rate at which error accumulation occurs.

Figure 9 illustrates the accumulation of error for both *float* and *posits*.
Examining the graph, it becomes evident that *float* accumulates error at a

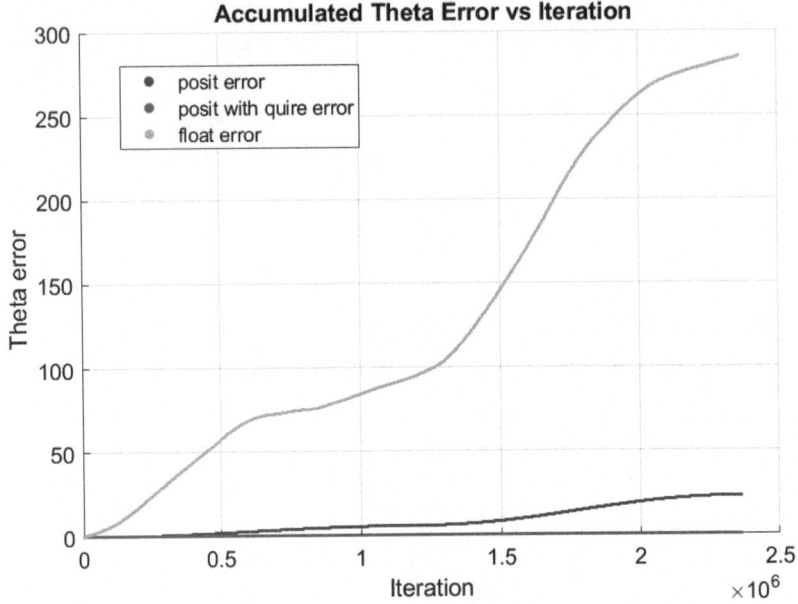

Fig. 9. Theta Error Accumulation.

significantly higher exponential rate compared to *posits*. Based on this observation, we can safely claim that *float* produces more errors than *posit* for this simulation. The analysis reveals significant differences in the number of instances for different numerical representations when considering an initial theta(θ) value of 90° and a time period of 2.368596 s, with a time step (Δt) of 10^{-6} s. When using *double* precision, there are 2368598 instances of t, which are considered as the expected count. *Float* precision results in 2363039 instances, missing 5557 values compared to the expected count. However, *posit* precision yields 2368538 instances, with only 58 time(t) values missing, indicating a substantial decrease in missing values compared to *float* precision. Furthermore, when employing *posit* with the *quire*, there are 2368597 instances, just one value fewer than those generated by *double* precision. These findings provide concrete evidence of the superiority of *posits* over *float* in simulations, as they demonstrate a more accurate representation of the time values with a significantly lower number of missing instances.

To further substantiate the claim of *posits*' exceptional performance in simulations, quantitative metrics are employed. Table 3 provides error details regarding theta(θ), omega(ω), and time(t) parameters for an initial angle of 90°, offering a comprehensive overview of the error in simulation parameters. Additionally, Fig. 10 illustrates the theta(θ) error for various initial angles, showcasing the advantages of *posit* over *float*. It becomes evident that the relatively low error exhibited by *posits* is not limited to a specific case but holds true for any

Table 3. Error metrics for θ, ω and t

	float	posit	posit quire
$\theta_{abs.mean\_error}$	0.00012039	9.7109e−06	6.3578e−09
θ_{max_error}	0.00039965	2.4267e−05	1.6983e−08
θ_{min_error}	4.6334e−11	5.5289e−14	2.2204e−16
$\omega_{abs.mean\_error}$	0.00042985	2.7219e−05	1.7983e−08
ω_{max_error}	0.0014318	7.4463e−05	4.8641e−08
ω_{min_error}	0	0	0
$t_{abs.mean\_error}$	0.01523	0.0011431	3.0211e−09
t_{max_error}	0.054105	0.0029531	1.3389e−08
t_{min_error}	0	0	0

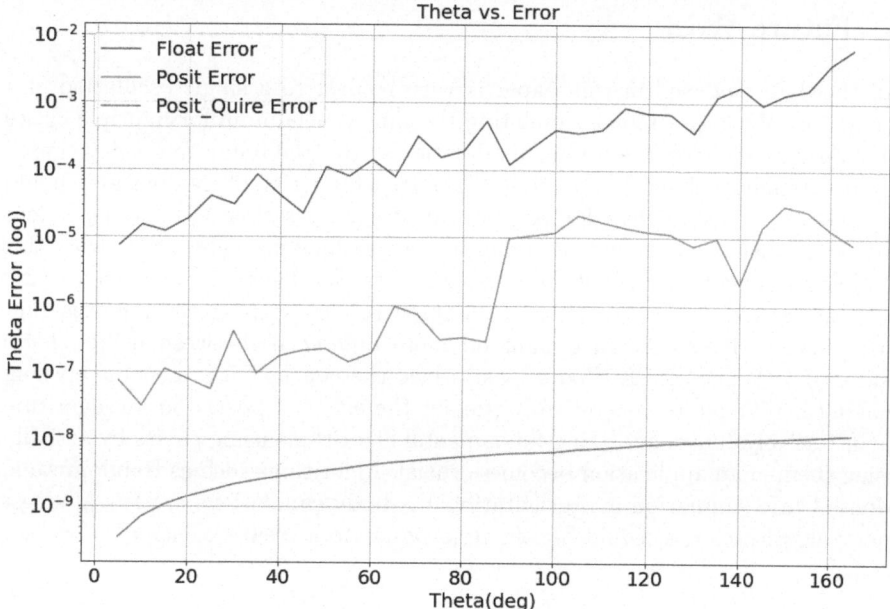

Fig. 10. Theta Error vs Theta Plot

theta(θ) value in general. We should especially note the reduction in error when *quire* is employed with respect to using *posits* without it.

7 Conclusion

After observing the performance of *posits*, particularly when combined with *quire*, in comparison to *float*, it is reasonable to consider them as potential replacements for *double* precision, when the switch in data type is due to a requirement of slightly more precision than what *float* can provide. *Posits* offer a balance between accuracy and memory efficiency, making them suitable for scenarios where extreme precision, such as 16 decimal places provided by *double*, is not necessary but the accuracy provided by *float* is not sufficient. This utilizing only 50% of the existing memory is sufficient for performing computations and obtaining results with accuracy greater than *float* and closer to that of *double*. Moreover, the reduced memory allows for fitting twice as many values in memory at a time compared to using *double* precision. This reduction in memory access delays should theoretically result in significantly improved overall computation time (around 2x), as it enables processing of twice the data points per second, this is of course assuming that posits are as fast as floats. This also hints at a possibility that *posit*<64, 3> might be a potential replacement for 128 bit IEEE floating point giving us the equivalent of one turn of Moore's law improvement.

8   Future Work

All the data analysed in this paper is with respect to a single oscillation of a pendulum. We suspect that simulating the simple pendulum for multiple cycles may reveal an inherent damping/modulation of the pendulum frequency caused due to rounding errors in the time variable. This effect of modulation should however be relatively less for *posits* than *floats*. As time is an independent variable, other time dependent simulation algorithms may also suffer from this modulation and thus benefit from using *posits*. We have also avoided varying parameters such as length (l) and damping (q) for keeping the analysis simple. Varying them may shed light on more differences between using *floats* and *posits*. Unlike IEEE *floats*, *posits* lack the concept of a machine epsilon, making it difficult to theoretically predict the effect of *posits* on an algorithm quantitatively. Thus, analysing fundamental algorithms using *posits* by actually using them in an application becomes crucial. Algorithms such as Random walk, Monte Carlo simulation and gravitational N-body simulations are few amongst many algorithms that could benefit from *posits* to a great extent.

References

1. Gustafson, J.L., Yonemoto, I.T.: Beating floating point at its own game: posit arithmetic. Supercomput. Front. Innov. **4**(2), 71–86 (2017)
2. Murillo, R., Del Barrio, A.A., Botella, G.: Deep PeNSieve: a deep learning framework based on the posit number system. Digit. Sig. Process. **102**, 102762 (2020)
3. Posit Working Group: Posit Standard Documentation, Release 3.2-draft (2018). https://posithub.org/
4. Brehmer, J.: Simulation-based inference in particle physics. Nat. Rev. Phys. **3**(5), 305 (2021)
5. Zhang, X., et al.: Digital quantum simulation of Floquet symmetry-protected topological phases. Nat. **607**(7919), 468–473 (2022)
6. Braconnot, P., et al.: Evaluation of climate models using palaeoclimatic data. Nat. Clim. Chang. **2**(6), 417–424 (2012)

7. Kochkov, D., Smith, J.A., Alieva, A., Wang, Q., Brenner, M.P., Hoyer, S.: Machine learning-accelerated computational fluid dynamics. Proc. Natl. Acad. Sci. **118**(21), e2101784118 (2021)

8. Kumar, R., Skinner, J.L.: Water simulation model with explicit three-molecule interactions. J. Phys. Chem. B **112**(28), 8311–8318 (2008)

9. Conzelmann, G., Boyd, G., Koritarov, V., Veselka, T.: Multi-agent power market simulation using EMCAS. In: IEEE Power Engineering Society General Meeting, pp. 2829–2834. IEEE (2005)

10. Asgari, M., Yang, W., Lindsay, J., Tolson, B., Dehnavi, M.M.: A review of parallel computing applications in calibrating watershed hydrologic models. Environ. Modell. Softw., 105370 (2022)

11. Maksum, Y., et al.: Computational acceleration of topology optimization using parallel computing and machine learning methods-analysis of research trends. J. Ind. Inf. Integr. **28**, 100352 (2022)

12. Naik, D., Ramesh, D., Gandomi, A.H., Gorojanam, N.B.: Parallel and distributed paradigms for community detection in social networks: a methodological review. Exp. Syst. Appl. **187**, 115956 (2022)

13. McKee, S.A.: Reflections on the memory wall. In: Proceedings of the 1st Conference on Computing Frontiers, p. 162 (2004)

14. Wulf, W.A., McKee, S.A.: Hitting the memory wall: implications of the obvious. ACM SIGARCH Comput. Archit. News **23**(1), 20–24 (1995)

15. Buoncristiani, N., Shah, S., Donofrio, D., Shalf, J.: Evaluating the numerical stability of posit arithmetic. In: 2020 IEEE International Parallel and Distributed Processing Symposium (IPDPS), pp. 612–621. IEEE (2020)

16. De Dinechin, F., Forget, L., Muller, J.-M., Uguen, Y.: Posits: the good, the bad and the ugly. In: Proceedings of the Conference for Next Generation Arithmetic, pp. 1–10 (2019)

17. Lindstrom, P., Lloyd, S., Hittinger, J.: Universal coding of the reals: alternatives to IEEE floating point. In: Proceedings of the Conference for Next Generation Arithmetic, pp. 1–14 (2018)

18. Klöwer, M., Düben, P.D., Palmer, T.N.: Posits as an alternative to floats for weather and climate models. In: CoNGA 2019 (2019)

19. Lim, J.P., Shachnai, M., Nagarakatte, S.: Approximating trigonometric functions for posits using the CORDIC method. In: Proceedings of the 17th ACM International Conference on Computing Frontiers, pp. 19–28 (2020)

20. Allen, E., Burns, J., Gilliam, D., Hill, J., Shubov, V.: The impact of finite precision arithmetic and sensitivity on the numerical solution of partial differential equations. Math. Comput. Model. **35**(11–12), 1165–1195 (2002)

21. Bailey, D.H., Barrio, R., Borwein, J.M.: High-precision computation: mathematical physics and dynamics. Appl. Math. Comput. **218**(20), 10106–10121 (2012)

22. Kettimuthu, R., Liu, Z., Wheeler, D., Foster, I., Heitmann, K., Cappello, F.: Transferring a petabyte in a day. Futur. Gener. Comput. Syst. **88**, 191–198 (2018)

23. Baker, G.L.: The Pendulum A Case Study in Physics. Oxford University Press, USA (2005)

24. Giordano, N.J., Nakanishi, H. (eds.): Computational Physics, 2nd edn. Pearson-/Prentice Hall (2006)

25. Goldberg, D.: What every computer scientist should know about floating-point arithmetic. ACM Comput. Surv. **23**(1), 5–48 (1991)

26. https://posithub.org/docs/posit_standard-2.pdf

27. https://calligotech.com/uttunga/

28. Klöwer, M., Coveney, P.V., Paxton, E.A., et al.: Periodic orbits in chaotic systems simulated at low precision. Sci. Rep. **13**, 11410 (2023). https://doi.org/10.1038/s41598-023-37004-4
29. Murillo, R., Del Barrio, A.A., Botella, G.: The effects of numerical precision in scientific applications (2022)

Author Index

M. Michalewicz et al. (Eds.): CoNGA 2024, LNCS 14666, p. 125, 2024.
https://doi.org/10.1007/978-3-031-72709-2